D0225715

The Biology of Millipedes

Cover illustration: An undescribed siphonophorid from Thailand. This order includes the millipede with the greatest number of legs so far recorded. The specimen illustrated is 5 cm long and has 650 legs. Redrawn from: Enghoff, H. (1984). Tusind ben? *Dyr i natur ag museum 1984*, **1,** 29. Used by kind permission of Henrik Enghoff. Original drawing by R. Neilsen.

The Biology of Millipedes

STEPHEN P. HOPKIN

Department of Pure and Applied Zoology
School of Animal and Microbial Sciences
University of Reading

and

HELEN J. READ

Ecologist for the Corporation of
London at Burnham Beeches

Oxford New York Tokyo
OXFORD UNIVERSITY PRESS
1992

Oxford University Press, Walton Street, Oxford OX2 6DP
Oxford New York Toronto
Delhi Bombay Calcutta Madras Karachi
Petaling Jaya Singapore Hong Kong Tokyo
Nairobi Dar es Salaam Cape Town
Melbourne Auckland
and associated companies in
Berlin Ibadan

Oxford is a trade mark of Oxford University Press

Published in the United States
by Oxford University Press, New York

A catalogue record for this book is available from the British Library

Library of Congress Cataloging in Publication Data
Hopkin, Stephen P.
The biology of millipedes / Stephen P. Hopkin, Helen J. Read.
Includes bibliographical references and index.
1. Millipedes. I. Read, Helen J. II. Title.
QL449.6.H67 1992 595.6'1—dc20 91–40375

ISBN 0–19–857699–4

Typeset by Footnote Graphics, Warminster, Wiltshire
Printed in Great Britain by
St. Edmundsbury Press, Bury St. Edmunds, Suffolk

To J. Gordon Blower

Preface

In this book we have attempted to present an up to date account of all aspects of the biology of millipedes for those studying the group for the first time and specialists alike. Until now, millipedes have been the only major group of terrestrial invertebrates for which there has been no general introductory text in English. We hope to show that millipedes, far from being a 'relatively uniform and unspectacular lot' (Eisner *et al.* 1978), are fascinating creatures, at least as worthy of study as their six-legged relatives. Even Newport (1841) bemoaned the preoccupation of naturalists with insects to the detriment of other arthropods!

We estimate that there are some 1200 to 1500 references in the literature that deal with non-taxonomic aspects of millipedes; about half of these are referred to in this book. Taxonomic descriptions of species run into many thousands. Since this is an account of the *biology* of millipedes (rather than their classification) we have not attempted to quote from more than a few of these publications. Recent reviews on classification by authors such as Enghoff (1984*a*), Hoffman (1979, 1982), and Jeekel (1970) should be consulted for further information. For comprehensive bibliographies of the earlier literature, specialists should refer to the works of Attems (1926), Blower (1985), Brade-Birks (1974), Brolemann (1935), Demange (1981), Hoffman (1979, 1982), Koch (1863), Newport (1841, 1843), Schömann (1956), Schubart (1934), Seifert (1932), Silvestri (1903), and Verhoeff (1926–1932). The series of papers by Manton (1954, 1956, 1958, 1964, 1966, 1973, 1974) contains much important background information. This she summarized in a book which is now regarded as a classic (Manton 1977).

In the UK, knowledge of the distribution of millipedes is excellent thanks to fieldwork conducted by members of the British Myriapod Group. Records of distribution have been collated by the Biological Records Centre and published recently in 'atlas' form (British Myriapod Group 1988). Modern checklists for other countries are beginning to appear (e.g. Geoffroy 1990) as are maps of distribution throughout Europe (Kime 1990*a*). However, detailed knowledge of the distribution of tropical species is almost non-existent.

Myriapodologists (those who study millipedes, centipedes, pauropods, and symphylids) and onychophorologists (*Peripatus* etc.) are well-served by a small but dedicated team of French zoologists who run the Centre International de Myriapodologie (CIM) at the Muséum d'Histoire Naturelle in Paris. An annual report has been produced since 1968 that lists all myriapodological publications of the previous year together with a list of researchers throughout the world who are currently studying the groups. To register with the CIM, write to:

Centre International de Myriapodologie
Muséum National d'Histoire Naturelle
Laboratoire de Zoologie (Arthropodes)
61 rue de Buffon
F-75005 Paris
France

Our task of searching out millipede references in a very diverse literature was made much easier by having so many of them collected together in the CIM annual reports.

Myriapodologists throughout the world meet regularly at the International Congresses. These have been held in Paris (1968), Manchester (1972), Hamburg (1975), Gargnano (1978), Radford (1981), Amsterdam (1984), Vittorio Veneto (1987), and Innsbruck (1990). The 9th Congress is planned for Paris in 1993 at which the 25th anniversary of CIM will be celebrated. The proceedings of most of these congresses have been published (Blower 1974*a*; Kraus 1978*a*; Camatini 1979; Jeekel 1985*a*; Minelli 1990) and all contain a wealth of information, some very specialized, on myriapods and onychoporans.

We hope that those who have some knowledge of millipedes will feel that we have done the group justice and that those starting at the beginning will feel inspired to learn more.

Reading
November 1991

S.P.H.
H.J.R.

Acknowledgements

We would like to express our sincere thanks to J. Gordon Blower for kindling our initial interest in millipedes and for his help and patience over the years. We are also extremely grateful to Henrik Enghoff for providing continuous enthusiasm and ideas.

Our task has been made considerably lighter as a result of help from the British Myriapod Group and the Centre International de Myriapodologie in Paris. The yearly meetings of the former never cease to provide stimulation (and amusement!) and the tri-annual International Congresses of the latter are a welcome opportunity to discuss myriapodological problems with our colleagues throughout the world. To all these people, we extend our sincere thanks for their help and support. We are especially grateful to those researchers who (without exception) have provided permissions to reproduce figures and tables from their publications, in many cases lending original diagrams and photographs. Where originals were not available, Reading University Photographic Services performed an excellent job in copying the figures.

During the preparation of this book, we have worked in the Reading University Library and the Zoologisk Museum, Copenhagen. We are grateful for the support offered by these institutions and to our colleagues therein. One of us (H.J.R.) would also like to thank the Corporation of London (especially Mark Frater) and Sue and John Davie for their understanding and help.

Contents

1

General introduction

1.1 What are millipedes?

Millipedes do not as their name suggests have a thousand legs. Indeed the 'world champion' *Illacme plenipes* has only(!) 375 pairs (Enghoff 1990), and most have fewer than fifty pairs. Nevertheless, it is the presence of so many legs arranged in two pairs per body segment which is their most characteristic feature. The presence of these **diplosegments** separates the Class Diplopoda (millipedes–sometimes millepedes or millipeds) from the three other classes of Myriapoda, the Chilopoda (centipedes), Symphyla, and Pauropoda, all of which have only one pair of legs per segment (Fig. 1.1.). It has been estimated that there are approximately 10,000 described species of millipedes in the world (Hoffman 1979; IUCN 1983). These range in size from as little as 2 mm to 30 cm in length.

The Class Diplopoda has been separated traditionally into two subclasses. The Subclass Penicillata (or Pselaphognatha) are a small group of quite common and widespread tiny millipedes known as 'bristly' millipedes. These are found usually under bark, have an uncalcified cuticle, and are covered with numerous serrated setae that give them a 'brush-like' appearance (Figs. 1.2, 6.5). The Subclass Chilognatha contains the vast majority of millipedes. These have a calcified cuticle, are generally long and thin with numerous legs, and live in or near the surface of the soil.

The 'doubling up' of leg-pairs enables millipedes to exert a considerable forward thrust. This, together with the calcified head capsule and leverage provided by the labrum (Fechter 1961), allows millipedes to force their way between fibres of rotting wood, and closely packed soil particles, and to enter microhabitats that may be unavailable to other terrestrial invertebrates. Indeed Manton (1977) remarked that 'a marked ability to push, using the motive forces of the legs, is as diagnostic of the Diplopoda as is the possession of diplosegments'.

1.2 Habitats and lifestyle

The vast majority of millipedes are detritivores and prefer to eat decaying plant material rather than living vegetation. Apparently there is only one order, the Callipodida, in which some members prefer food of animal

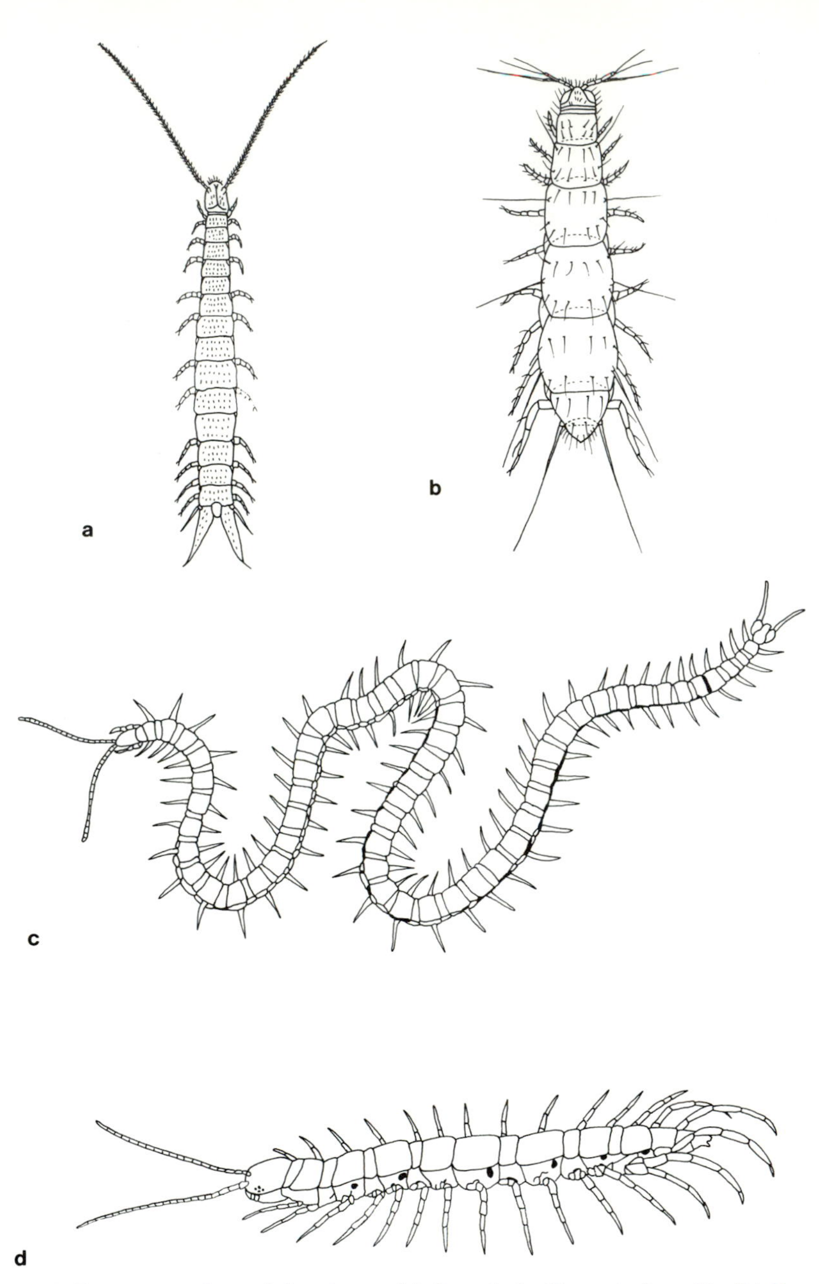

Fig. 1.1 Representatives of the classes (a) Symphyla (5 mm in length), (b) Pauropoda (2 mm in length), (c, d) Chilopoda showing the two main ecomorphological types. (c) *Necrophloeophagus longicornis* (burrowing type, 7 cm in length), (d) *Lithobius forficatus* (running type, 3 cm in length). The possible phylogenetic relationships between these classes and the Diplopoda are shown in Figs 2.3 and 2.4. After various authors.

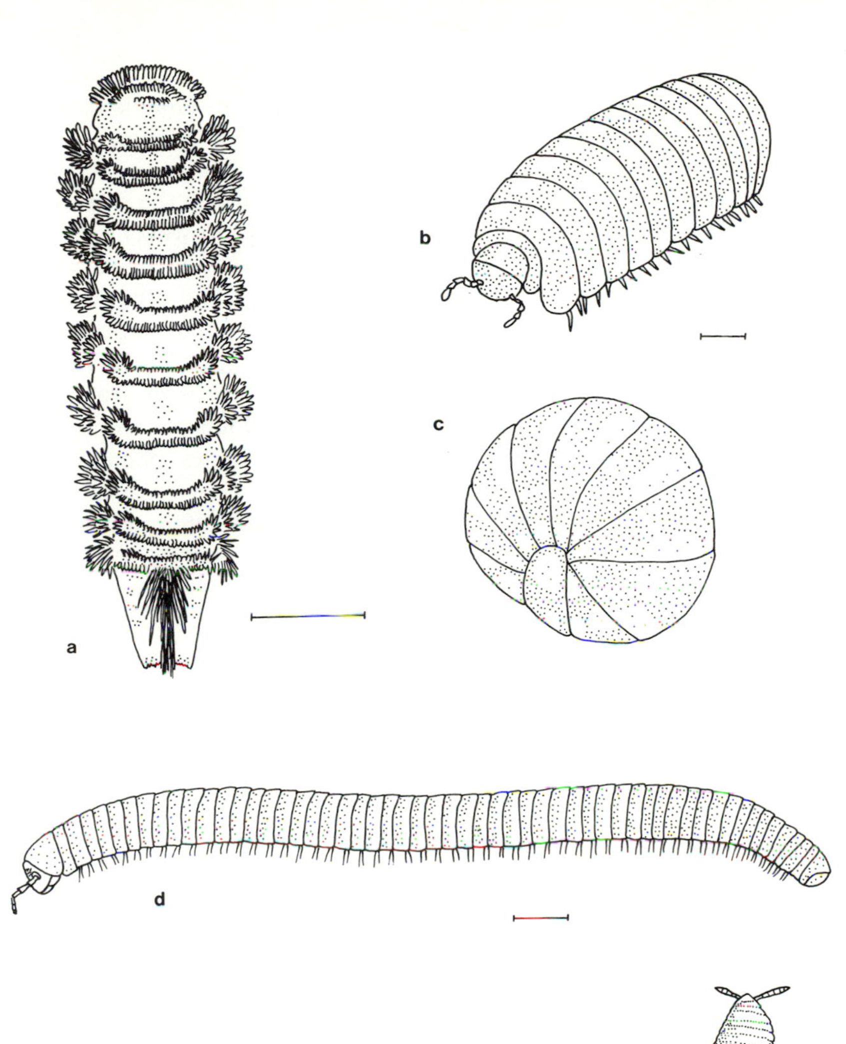

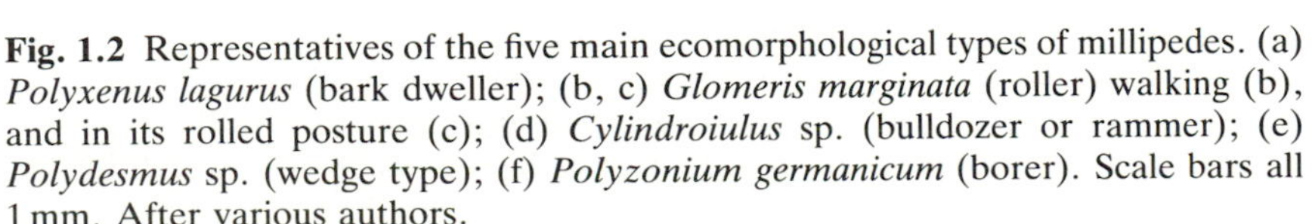

Fig. 1.2 Representatives of the five main ecomorphological types of millipedes. (a) *Polyxenus lagurus* (bark dweller); (b, c) *Glomeris marginata* (roller) walking (b), and in its rolled posture (c); (d) *Cylindroiulus* sp. (bulldozer or rammer); (e) *Polydesmus* sp. (wedge type); (f) *Polyzonium germanicum* (borer). Scale bars all 1 mm. After various authors.

rather than of plant origin (Hoffman and Payne 1969). Several species of millipedes can be found associated occasionally with the remains of dead animals but it is not always clear whether they were feeding on the decaying tissues, or were using the corpses as merely a moist shelter-site.

Millipedes are found usually under leaf litter or stones on the surface of the soil, or below the surface among soil particles. However, several species may climb trees and others can be quite common in very dry environments such as deserts. Millipedes, along with many other arthropods, may associate with ants. The ants provide protection from predators and the millipedes help to clean up detritus from the nests of their hosts (Murakami 1965; Rettenmeyer 1962). Some species live with army ants and can be found walking along trails of raiding columns (Hölldobler and Wilson 1990). Others occur in the nests of termites (Chamberlin 1923).

The development of millipedes has attracted considerable interest over the years. In general, they take one or two years to mature into adults although in some species it may take longer. The length of this development time can be correlated with type of habitat. Of special interest to diplopodologists is the phenomenon of **periodomorphosis**. This occurs in several species of the Order Julida. Adult millipedes that exhibit periodomorphosis moult from a sexually mature individual to a non-sexual stage. The reasons behind this phenomenon are not completely understood but are discussed at length in Section 8.5.4.

Millipedes are long-lived animals in comparison to most other terrestrial arthropods. The common pill millipede *Glomeris marginata*, for example, takes several years to mature and can live for 11 years (Carrel 1990). Their longevity may be related to the poor quality of food consumed (Blower 1985).

The limited dispersal powers of millipedes has resulted in a high degree of speciation, and the evolution of a large number of endemic species in many areas which have very restricted ranges. Several of these species are particularly vulnerable to small scale environmental change and may be threatened by Man's activities.

1.3 Ecological importance

Most ecologists now accept that the main role of detritivorous invertebrates in enhancing decomposition of dead plant material is to stimulate microbial activity (Price 1988). The fragments of leaf that are voided in the faeces have a greater surface area available for colonization by bacteria and fungi than the original leaf material. Direct chemical decomposition of leaf litter and wood by soil invertebrates is small in comparison to that of microbes. However, in the absence of the comminuting activity of soil invertebrates (for example, when leaf litter is left in the field in small mesh litter bags, or pollution reduces their populations), decomposition is greatly reduced (Anderson 1988; Heath *et al.* 1966; Hopkin *et al.* 1985; Swift *et al.* 1979).

There are few habitats in which millipedes are responsible for ingesting more than 5–10 per cent of the annual leaf litter fall; earthworms are usually the dominant detritivores (Bocock 1963; Van der Drift 1975). However, when earthworms are scarce, millipedes may occur at densities of several hundred per square metre and consume 25 per cent of the annual litter fall (Blower 1970*a*, 1974*b*).

The role of millipedes in decomposition processes is covered more extensively in Section 10.2.

1.4 Millipedes and Man

Millipedes do not impinge on the activities of humans to a major extent. However, there have been instances of millipedes being of local nuisance value (see Section 10.5). In Japan, for example, migrating 'swarms' of millipedes may be so large that they have earned the name 'train millipedes' (Fig. 1.3). Their squashed remains on railway tracks prevent the wheels from gaining a purchase on the rails (Niijima and Shinohara 1988; Shinohara 1981).

The 'spotted-snake' millipede *Blaniulus guttulatus* sometimes reaches pest proportions (MAFF 1984) and may cause significant economic damage to root crops (Blower 1985). In Australia, the julid *Ommatoiulus moreleti* is a pest in some areas where it enters houses in large numbers (Fig. 10.5; Baker 1985*a*; McKillup and Bailey 1990). The species was introduced to Australia from Portugal and its rapid spread may be due to the lack of natural control from predators and parasites in its new home (see Section 10.4).

Invertebrate zoologists who study groups of little apparent economic importance may long for someone to discover that their animal will provide a cure for AIDS, or some other serious disease that afflicts the human population. A paper in the *Journal of Traditional Chinese Medicine* suggests that millipedes may be worth more than a cursory glance in this respect (Tinliang *et al.* 1981). Extracts of the millipede *Spirobolus bungii*, when applied to tissue cultures of malignant tissues, arrested mitosis of tumour cells for six hours. This is clearly worthy of further investigation since defensive secretions of other species in the genus *Spirobolus* have a powerful necrotic effect on human skin (see below).

One of the most interesting aspects of millipede biology (and one which will be covered in detail in Section 9.3) is the secretion of defensive chemicals. While these would have evolved to defend millipedes against attack by animals other than humans, contact with these secretions may result in considerable individual discomfort (Radford 1975). The following passage taken from an article by Burtt (1938) illustrates this point admirably.

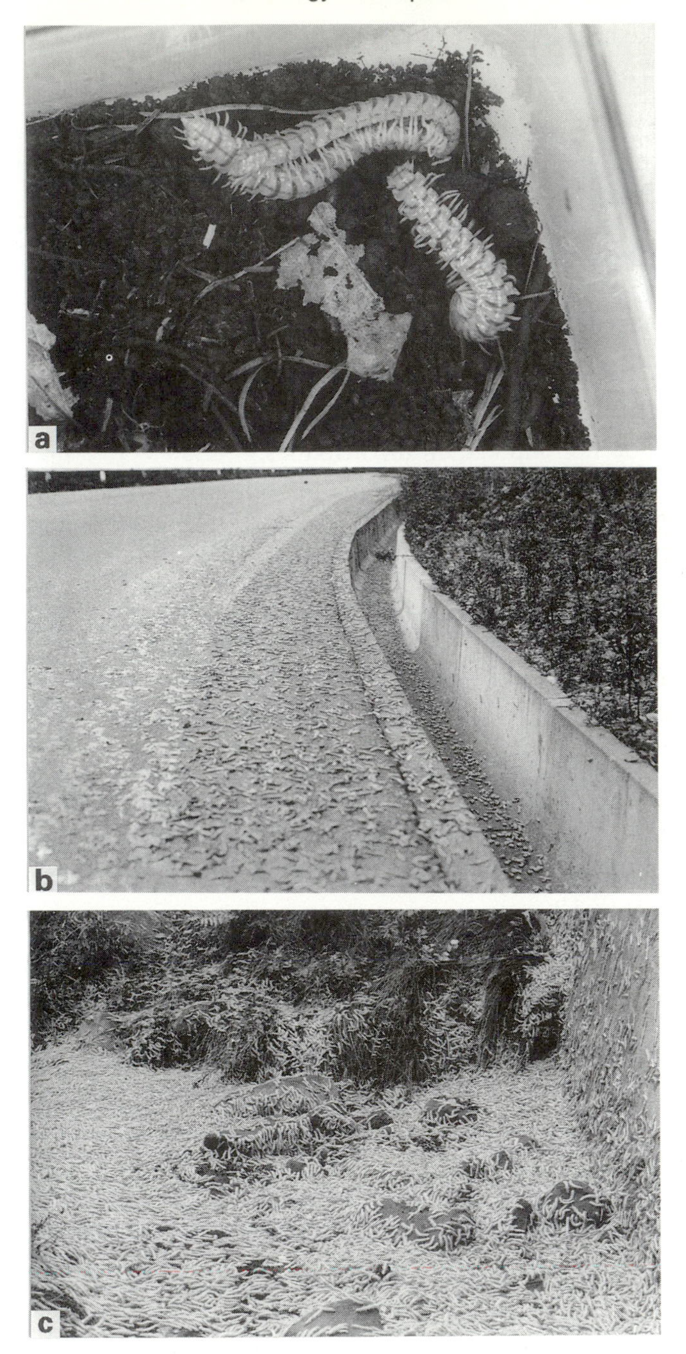
a
b
c

Fig. 1.3 The Japanese 'train millipede' *Parafontaria laminata laminata*. (a) Mating pairs. (b) and (c) Outbreaks of 'swarms' of this species at Nobeyama, Nagano Prefecture during September 1984. Reproduced from Niijima and Shinohara (1988) by kind permission of the authors and the Ecological Society of Japan.

While working at Sigi, 1,500 feet below the East African Agricultural Research Station at Amani, Tanganyika Territory, one evening in June, I came across one of the giant black millipedes–a species of *Spirobolus*–which are fairly common in this region of tropical evergreen rain-forest. It was an unusually large specimen, fully twelve inches in length. Having no box capable of holding it, I buttoned it up in my hip pocket and continued my work for an hour or so.

I felt the millipede moving about in my pocket and noticed that I was becoming rather sore in that neighbourhood, but paid little attention to it. However, whilst bathing shortly afterwards I was surprised to find that my skin had become completely blackened over an area of about nine square inches, with further red inflammation spreading rapidly down my thigh. Four days later all this blackened skin sloughed away, leaving a raw wound. This happened in June 1937; at the end of August 1938 the site of the injury is still visible.

I have since examined millipedes of the same and other species on several occasions, and noticed that, when one is molested by being turned about in the fingers, small drops of liquid are exuded from pores, one on the side of each segment. The liquid is rich yellow brown in colour and stains the fingers like iodine; it has a characteristic pungent odour recalling that of nitrogen peroxide, but is neutral to litmus. The fumes cause marked watering of the eyes. Mere contact of the skin with the tough skin of the fingers produces no injurious symptoms, although when some was rubbed on to the skin of the leg, smarting was experienced and the skin eventually became hard and scaly.

1.5 Millipedes as experimental models

Many species of millipedes are large and easy to keep in the laboratory. Marcus *et al.* (1987) suggested that *Triaenostreptus* was an ideal model for studying arthropod physiology because it grows up to 30 cm in length and repeated blood samples can be taken from it. Millipedes have also been used to study the mechanisms behind orientation (Mittelstaedt *et al.* 1979) and results have been obtained which have helped our understanding of the processes involved in other invertebrates.

A research area which is receiving much attention at present is that of the genetic control of segmentation in arthropods. Most of the work has focused on embryos of *Drosophila*. However, Minelli and Bortoletto (1988) have pointed out that myriapods provide excellent experimental models in which to examine segmentation since they have so many! Elucidation of the molecular mechanisms that control development of segment number in millipedes (and centipedes) would provide information that could be applied to other arthropods in which segment number is apparently fixed.

2

Taxonomy, evolution, and zoogeography

2.1 Synopsis of orders of the Class Diplopoda

The following descriptions are intended to illustrate the range of forms found in the Class Diplopoda and to provide an introduction to the main groups referred to in the text. More detailed descriptions of each family can be found in Hoffman's (1982) very useful guide. Classification is based on Enghoff (1984*a*) which is reproduced in Table 2.1

2.1.1 Subclass Penicillata

2.1.1.1 Order Polyxenida

The polyxenid or 'bristly' millipedes are rather different from all other millipedes. They are very small (<4 mm in length), have only 11 to 13 segments, and are aften found in dry places. The body wall is soft and is not impregnated with calcium salts (unlike all other millipedes). The body is covered with tufts of serrated bristles (Figs 1.2a, 3.14b, 6.5). During mating, the male draws the female over a package of sperm (Fig. 7.4). There are no specialized appendages for sperm transfer.

2.1.2 Subclass Chilognatha; Infraclass Pentazonia

The Pentazonia includes the pill millipedes, which are able to roll into complete spheres. In this group, the last one or two pairs of legs in the male are enlarged to form telopods or claspers which assist in sperm transfer. *Penta*zonia refers to the five cuticular components that make up each body ring, i.e. a tergal arch, two pleurites, and two sternites (Fig. 3.5).

2.1.2.1 Order Glomeridesmida

These are small, rather primitive-looking blind millipedes found in tropical regions. The body is rather flattened and consists of 22 segments. These animals are unable to roll into balls and are possibly similar to the ancestral millipede (Fig. 2.6).

2.1.2.2 Order Sphaerotheriida

The so-called giant pill millipedes (up to 10 cm in length) are found predominantly in the Southern Hemisphere. They have 13 segments and can roll into complete spheres. In some species, the male can stridulate,

Table 2.1 A cladistic classification of the Class Diplopoda (after Enghoff 1984*a*).

Class Diplopoda

Subclass Penicillata
Order Polyxenida

Subclass Chilognatha
Infraclass Pentazonia
Order Glomeridesmida
Order Sphaerotheriida
Order Glomerida
Infraclass Helminthomorpha
Helminthomorpha incertae sedis: Order Siphoniulida
Subterclass Colobognatha
Order Platydesmida
Order Siphonophorida
Order Polyzoniida
Subterclass Eugnatha
Superorder Nematophora
Order Stemmiulida
Order Callipodida
Order Chordeumatida
Superorder Merocheta
Order Polydesmida
Superorder Juliformia
Order Spirobolida
Order Spirostreptida
Order Julida

producing a sound by rubbing the last legs against the sides of the last tergite.

2.1.2.3 Order Glomerida

The Glomerida are the pill millipedes of the Northern Hemisphere which are generally rather smaller in size (up to 2 cm in length) than their southern relatives. The most common species in Britain is *Glomeris marginata* (Figs 1.2 b,c, 3.14a) which has been the subject of numerous ecological and physiological studies. Like the Sphaerotheriida, they have 13 segments but the second and third tergites are fused to form a very large plate. They are also able to roll into a ball and some of the males stridulate. Several related species in continental Europe bear brightly coloured spots which form an attractive contrast with the jet-black ground colour of the cuticle. These millipedes are quite resistant to desiccation and are often seen walking over roadside kerbstones under the full glare of the Mediterranean sun.

2.1.3 Infraclass Helminthomorpha

The remaining millipedes are more elongate in shape. The (sub) cylindrical forms, including the large spirobolids, are how most people perceive millipedes. Some or all of anterior leg-pairs 6 to 8 in males are modified to form intromittent organs (gonopods) for the transfer of sperm to females. The penises of males and the vulvae of females open more anteriorly, around leg-pairs 2 or 3.

2.1.3.1 Order Siphoniulida

This order is very poorly known (from females only), and is represented by just two species at present. Siphoniulids are tiny (<7 mm in length) with smooth cylindrical segments. Ocelli are absent but the head is drawn out into a long rostrum. The taxonomic position of this group is uncertain at present (Table 2.1). They have been found in Sumatra and Guatemala.

2.1.3.2 Order Platydesmida

These millipedes may reach 60 mm in length with up to 110 segments. They have a flat-backed appearance and are blind. The male gonopods are rather simple and leg-like. In some species, the male may brood the eggs until they hatch.

2.1.3.3 Order Siphonophorida

This order has the distinction of including the millipede with the most legs (*Illacme plenipes*, with up to 375 pairs). Several species have in the region of 180 to 190 segments and there is also a wide intraspecific variation. Siphonophorids tend to be thin as well as long and have rather simple gonopods. The head is also characteristic as it lacks eyes and the mouth parts are drawn out into a long beak. The order is mainly tropical. Almost nothing is known about their biology.

2.1.3.4 Order Polyzoniida

These millipedes are dome-shaped in cross section (Figs 3.5a, 7.1). The gonopods are leg-like, the head is small and often pointed (Figs 1.2f, 3.13b), and the collum is enlarged. They are temperate and tropical.

2.1.3.5 Order Stemmiulida

Stemmiulids have segments which are rather higher than they are wide. The legs are slender and the anterior pair on segment 7 are modified into gonopods, the structure of which is unlike those of any other millipede order. The second pair of legs in the males is often modified into hook-like structures bearing setae. Stemmiulids are very active and some are able to jump (Fig. 3.15). They are confined to the tropics.

2.1.3.6 Order Callipodida

These millipedes are cylindrical in shape and range up to 10 cm in length with many segments. The last segment is reduced but bears a pair of spinnerets on the posterior margin. Most species have large numbers of eyes. The male gonopods are modifications of the posterior pair of legs on segment 7. Some callipodids are carnivorous (Hoffman and Payne 1969).

2.1.3.7 Order Chordeumatida

The millipedes in this order usually have 30 segments (total range 26 to 32). They may be cylindrical, subcylindrical, or with lateral projections (paranota) giving them a flat-backed appearance. On the whole, chordeumatids are found in the Northern Hemisphere, are usually rather long-legged and active, and are confined to damp places. Like the callipodids, they have spinnerets on the last segment.

2.1.3.8 Order Polydesmida

These are the true flat-backed millipedes. They have completely fused sclerites and, usually, strong lateral projections on the hind part of each segment (Figs 1.2e, 3.13a). Most have 20 segments (total range 18 to 21) and are blind. Gonopods are formed from the anterior leg pair on segment 7. There are a large number of families in the Polydesmida, which is the largest order of millipedes, containing more than 2700 species. Distributed worldwide, they range in length from 3 mm to 13 cm. Simonsen (1990) has dealt recently with the taxonomy of this group at family level.

The remaining three orders comprise the Superorder Juliformia (Table 2.1). These are heavily-calcified, cylindrical animals, often termed 'snake millipedes'. Apart from certain cave species which are blind, the majority possess ocelli.

2.1.3.9 Order Spirobolida

This order is distinguished by a pronounced suture that runs vertically down the front of the head. Both pairs of legs on the seventh segment of the male are modified into gonopods. The spirobolids are generally tropical species, some of which are very brightly coloured (Lewis 1984).

2.1.3.10 Order Spirostreptida

The Spirostreptida includes the largest millipedes known (up to 30 cm in length with as many as 90 segments), and some of the tiniest (<6 mm in length). The male gonopods consist of both leg-pairs of the seventh segment with the anterior pair being most active. The first pair of legs in the male is also highly modified into secondary sexual structures that are used during mating.

Spirostreptids and spirobolids are predominantly tropical, but are often found for sale in pet shops in more temperate climates.

2.1.3.11 Order Julida

These rather smaller cylindrical millipedes are found mainly in temperate regions (Figs 1.2d, 3.12). The male gonopods are formed from both pairs of legs on the seventh segment. In addition, the first and sometimes the second pair of legs in the male may be modified to form secondary sexual structures.

2.2 Fossil record

The myriapod fossil record is patchy. Certain formations are relatively rich whilst the intervening periods yield very little. Millipedes are better represented in the fossil record than centipedes, due partly to their calcified exoskeleton, and partly to their burrowing habits.

There are problems in interpreting millipede fossils. It is difficult to know if differences in numbers of legs and segments are true differences, or are due to the presence of immature stadia. Likewise, lack of fossilized gonopods hampers identification and may be due to primitive sperm transfer as seen in *Polyxenus* today, or again to immature animals. Almond (1985*a*) has pointed out that fossils are often assigned to the Diplopoda on the basis of absence of features rather than presence of particular characters.

The oldest myriapod find is of a Scutigeromorph centipede-type animal from the Upper Silurian just above the Ludlow bone bed (Jeram *et al.* 1990). *Kampecaris tuberculata* is perhaps the oldest definite millipede and was found in Silurian Old Red Sandstone. Previously described 'myriapods' (e.g. *Archidesmus*) have turned out to be plant material (Cloudsley-Thompson 1988).

The Devonian and Carboniferous periods saw the expansion of the Diplopoda and indeed the whole terrestrial decomposer community. The development of spines on the spores of various plants at this time may have been in response to increased grazing pressure by these animals.

Some early finds were considered to be aquatic forms due to their lack of spiracles, and presence in sediments laid down underwater. However, spiracles are tiny structures which may not be clearly preserved. An aquatic life-style was initially projected for a very interesting group of fossil 'myriapods', the Arthropleurida (Fig. 2.1). Originally described as a swimming centipede, *Arthropleura* from the Carboniferous, and its forerunner *Eoarthropleura* from the Devonian (Størmer 1976), have been assigned a variety of affinities. It seems that the segments of small *Arthropleura* may have been diplosegments (Almond 1985*b*) and that it may have been either a true myriapod, or have represented another group within the Uniramia (Hannibal 1986; Rolfe 1986).

Fig. 2.1 Reconstruction of *Arthropleura* (the detailed morphology of the head is unknown). Adult specimens probably exceeded 1 m in length. Drawing by Annemarie Burzynski reproduced from Briggs *et al.* (1984) by kind permission of the authors and the Palaeontological Society.

When alive, *Arthropleura* must certainly have been an impressive sight. Judging from fossilized remains and tracks, the animals were the largest terrestrial invertebrates ever to have walked on land. Almond (1985*b*) estimated a length of 1.8 metres and a breadth of 0.45 metres. *Arthropleura* resembled a giant polydesmid millipede with some 20 to 30 segments (Almond 1985*b*; Briggs *et al.* 1984). Vegetable matter formed a large part of its diet (Rolfe and Ingham 1967) and it probably lived in the lycopod forests (Rolfe 1986).

Between the Silurian and Carboniferous periods, the Archypolypoda occur in the fossil record. Classified as a separate class within the Myriapoda by Hoffman (1969), but as true diplopods by Burk and Kraus (Hannibal 1981), they were certainly very similar to millipedes. Fossils up to 30 cm in length have been found in North America. The group also includes *Acantherpestes* from Czechoslovakia which had four longitudinal rows of bifurcate spines. Archypolypods had large eyes, long legs and spines, and presumably were surface-living (Hannibal 1981).

In the Carboniferous period a range of definite millipede taxa can be found including *Pleurojulus*. This was probably the oldest Colobognath, similar to recent species of *Polyzonium* and *Platydesmida* (Dzik 1981). In addition, pill millipedes of the Order Amynilspedida appeared (Hannibal and Feldmann 1981). These were similar to the extant Glomerida and Sphaerotheriida and were up to 3 cm in length. Many bore spines like the archypolypods (Fig. 2.2).

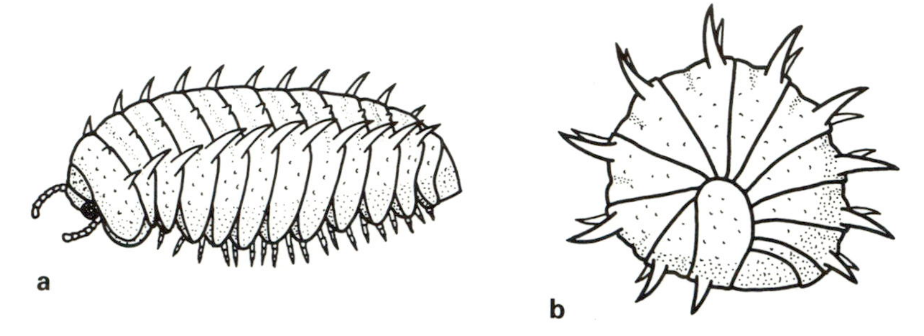

Fig. 2.2 Reconstruction of the spiny pill millipede *Amynilyspes* (of 3 cm in length) walking (a) and rolled into a ball (b). Reproduced from Hannibal (1984) by kind permission of the author and the Field Museum of Natural History, Chicago.

After the Carboniferous, the number of fossils declines. Some specimens from the late Palaeozoic resemble modern day forms such as *Isojulus*, which may have been a spirobolid (Kraus 1974). More spirobolids were described by Dzik (1975) from the late Cretaceous in Mongolia.

The next substantial collection of millipedes is from the Oligocene and are preserved in amber. By this stage, the species are very similar to modern day forms and include Polyxenida, Chordeumatida, Polydesmida, and Siphonophorida (Shear 1981). Hoffman (1969) gives a useful survey of myriapod orders and their known fossil record.

There is currently much debate on the faunistic composition of Palaeozoic terrestrial ecosystems (see Shear (1991) and Shear and Kukalova-Peck (1990) for recent discussions of this topic). It seems likely that as more material is discovered, millipedes will prove to have been a much more important component of the early decomposer community than has hitherto been realized.

2.3 Myriapod phylogeny

2.3.1 Introduction

The relationships between millipedes and other myriapods (and the insects) have been much discussed. Millipedes have been classified in a variety of positions in the past, for example as apterate insects by

Linnaeus, and as arachnids by Lamarck. A likeness to worms and Crustacea has also been suggested (Sinclair 1895).

The fossil record has not proved particularly useful, as yet, in evaluating prehistoric relationships. The result of this is a variety of theoretical approaches from different viewpoints. Recent workers, notably Dohle and Enghoff, have been exponents of cladistic principles. As a consequence, many of the more realistic phylogenies have a cladistic background and are represented as cladograms (e.g. Enghoff 1981). Brusca and Brusca (1990) provide a good 'beginners guide' to cladism, whereas Wiley (1981) gives a more detailed account of the principles involved.

2.3.2 Millipedes and other arthropods

As Enghoff (1984*a*) recorded, 'the monophyly of the Diplopoda has never been seriously disputed'. In other words it is likely that all members of the Class Diplopoda can be traced to a single ancestral form. Enghoff gave three reasons for this (**autapomorphies**—derived characters shared by all members of the group and not by other groups):

(1) having body segments fused into diplosegments;
(2) having aflagellate spermatozoa (Fig. 7.3); and
(3) having four (or more) large sensory cones on the apex of the distal segment of the antenna (Fig. 6.7).

Having established that the Diplopoda are derived from a common ancestor, it is rather more difficult to be as certain about the relationships within the Myriapoda and to prove monophyly of the group as a whole.

Early workers considered the Onychophora (velvet worms, including *Peripatus* and the like) to be ancestral to the Myriapoda (Fig. 2.3a,b). Later phylogenies placed the Onychophora as a sister group to various assemblages of arthropods (Fig. 2.3c). It seems most likely that the Onychophora are not the progenitors of the Myriapoda but that both groups originated from an ancestor with lobopods (Gupta 1979; Lanzavecchia and Camatini 1979).

There are two fundamentally different views on the phylogeny of the arthropods. The first held by Manton (1972) is that the Phylum is polyphyletic, i.e. a common ancestor gave rise to all the various forms (in early diagrams, Manton's lines did not connect basally; this was rationalized a little in 1972—Fig. 2.3e). This view of polyphyly is disputed by just about every other worker nowadays! Most researchers consider the Arthropoda to be a monophyletic group (e.g. Ax 1987; Dohle 1979).

It is becoming widely accepted that the diplopods are closest to the pauropods. These two classes have been jointly termed the Dignatha since both have two pairs of gnathal appendages (Dohle 1974*a*). There are various pieces of evidence that suggest that the Symphyla are the sister

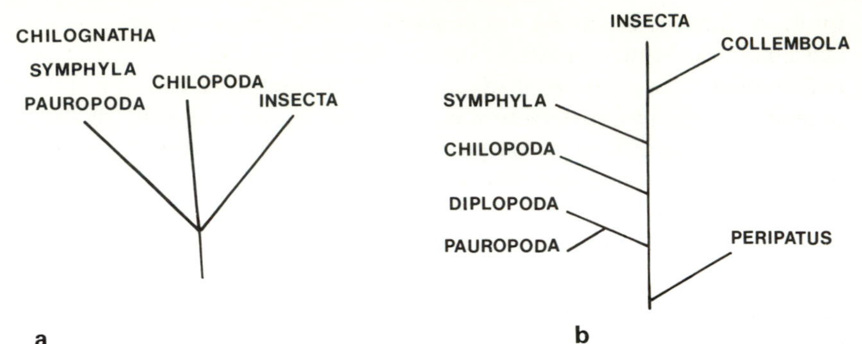

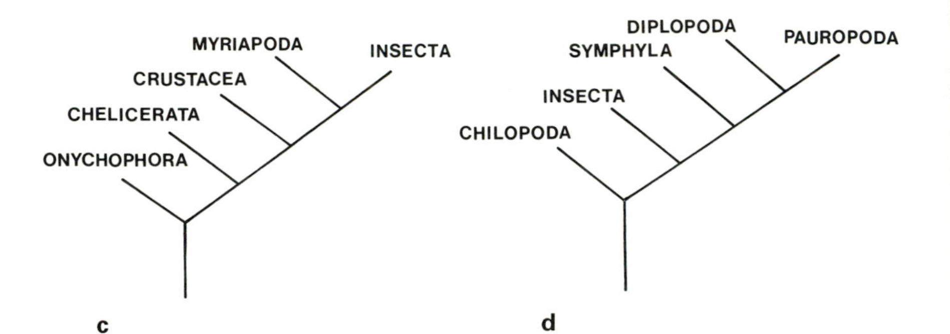

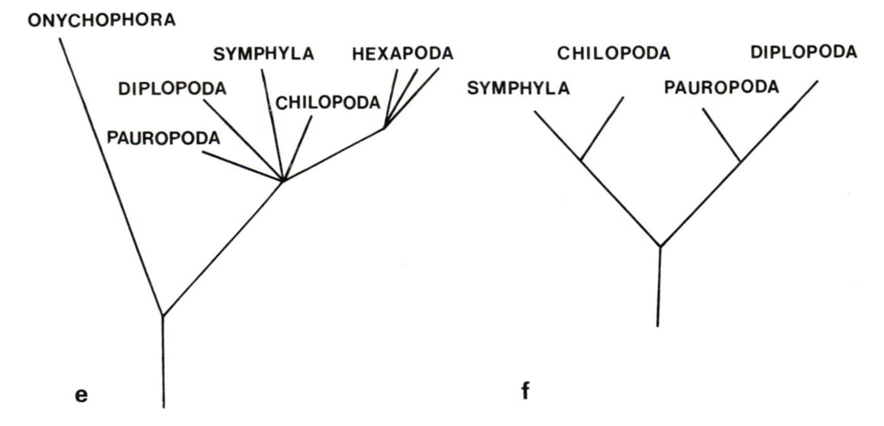

Fig. 2.3 Relationships between the Diplopoda and other arthropod groups as proposed by various authors. (a) Schmidt (1895), (b) Tiegs (1947), (c) Ax (1984), (d) Dohle (1965), (e) Manton (1972), (f) Boudreaux (1979).

group of the Dignatha but rather more speculation about the position of the Chilopoda. Dohle (1980, 1988) has discussed whether the insects or the chilopods were first to split off from the remaining myriapods, but was unable to come to a satisfactory conclusion. He suggested that there may be more reason to link the Progneata (Diplopoda, Pauropoda, and Symphylida) to the Insecta than to the Chilopoda.

Gupta (1979) has suggested that both the apterygote and pterygote insects originated from myriapodous ancestors similar to symphylids. Boudreaux (1979) came up with the alternative proposal that the chilopods are most closely related to the symphylids and that both form the sister group of the Dignatha (Fig. 2.3f).

It seems at the present that we are unable to resolve the phylogeny of the Myriapoda with any degree of certainty.

2.3.3 Elongation vs. contraction

One of the more hotly debated subjects is whether the ancestors of the millipedes had long bodies which reduced in number of segments during evolution (contraction theory), or whether they were short and became gradually longer (elongation theory). Among the early myriapodologists who addressed this problem, Brolemann and Attems were exponents of contraction whereas Verhoeff believed in elongation. This argument was discussed at length by Dohle (1988) who concluded, by cladistic reasoning, that the ancestral millipede probably had between 11 and 17 segments, thus agreeing with the theory of elongation (Fig. 2.4). It appears that elongation was probably the course taken by centipedes as well.

2.3.4 Phylogeny of Diplopoda

Hoffman (1979) gave a detailed historical account of the various classifications that have been proposed in the past for the Diplopoda. He concluded with a rather different version of his own, gave some phylogenetic details,

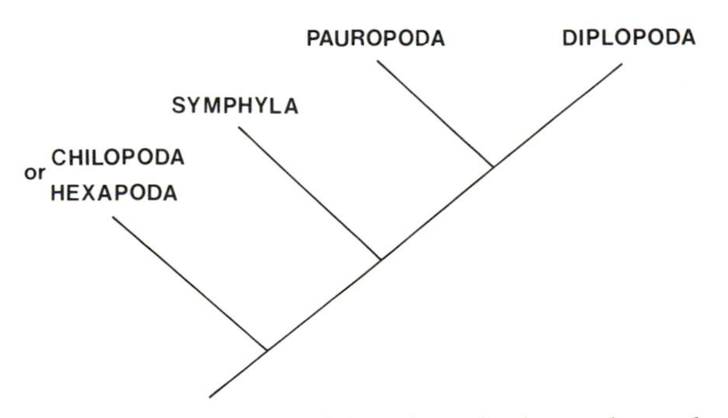

Fig. 2.4 Relationships between the Diplopoda and other arthropod groups based on Enghoff (1990).

and touched on the subject of evolution. In a more recent study however, Enghoff (1984*a*) produced a very detailed phylogeny based on cladistic principles. His resulting cladogram is shown in Fig. 2.5. Apomorphies (derived characters) are given for all groups (see Enghoff 1984*a* for details). The penicillate or bristly millipedes are often considered to be the most primitive and are distinguished from all remaining groups (the Chilognatha) by the presence of serrate setae in tufts (Fig. 6.5).

Enghoff (1990) has discussed the possible characters of the ground plan of the Chilognatha (i.e. their last common ancestor). The reconstruction is shown in Fig. 2.6 and is based on *Glomeridesmus barricolens* from Panama. The classification that resulted from the cladistic analysis is somewhat similar to many of the older classifications and is shown in Table 2.1. Names used in this book generally follow Hoffman (1979), an impressive and very useful publication that provides a very detailed classification. More recently, computer packages have been used to assist millipede

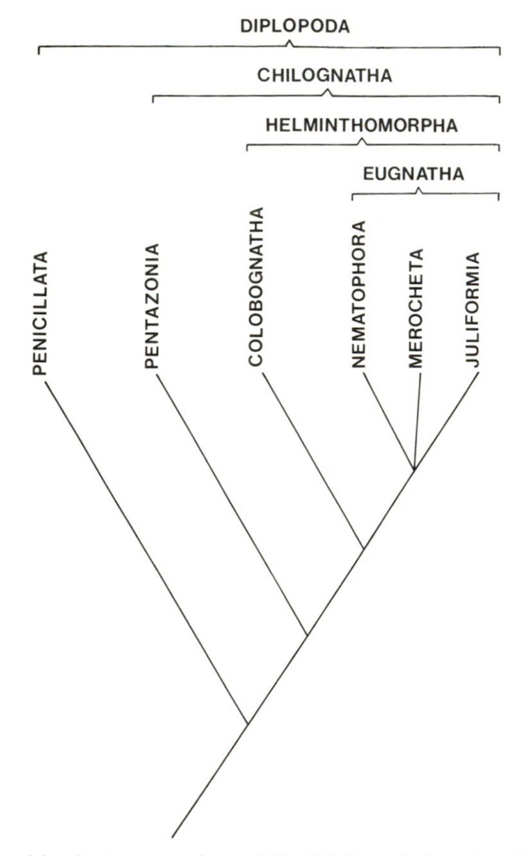

Fig. 2.5 Relationships between orders of the Diplopoda based on Enghoff (1984*a*). The Juliformia includes the Julida, Spirostreptida, and Spirobolida.

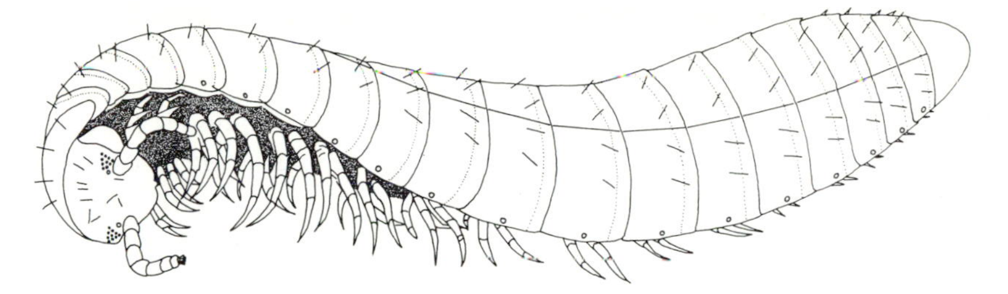

Fig. 2.6 Reconstruction of the ancestral chilognathan (based on a drawing of a female *Glomeridesmus barricolens* from Panama). Reproduced from Enghoff (1990) by kind permission of the author and E. J. Brill, Leiden.

taxonomists (Read 1990; Simonsen 1990) but we still have some way to go before the evolution of the myriapods is fully understood.

One other important work should be mentioned here, Jeekel's (1970) *Nomenclator generum et familiarum Diplopodorum*, a list of the genus and family group names in the Class Diplopoda. This is an extremely important work for diplopod taxonomists.

2.4 Zoogeography

The distribution of millipedes is influenced by a variety of factors acting at different levels. Global distribution is closely related to evolution and will be discussed here. Microhabitat factors are also important but these will be dealt with in Chapter 10.

There have been several attempts to examine the global distribution of millipedes and to relate these to plate tectonics and continental drift. Millipedes might be expected to follow any relevant patterns due to their poor ability to disperse. Today's patterns can be used to try to elucidate events of the past. Kraus (1978*b*) provided an interesting summary of zoogeography and plate tectonics in relation to millipede distributions.

Hoffman (1978) proposed continental drift to explain the distribution of present day Chelodesmidae (Polydesmida). This group is rather similar in both Africa and South America and the family must have pre-dated the separation of the two continents.

Jeekel (1985*b*) discussed the distribution of the julids and spirostreptids and suggested, from the Gondwanic distribution patterns, that evolution took place in the early Mesozoic or late Paleozoic. Thus these two groups have a southern origin. The holarctic distribution of the Julida resulted from northwards dispersion, followed by extinction in the Southern Hemisphere.

A further example is provided by the Southern Hemisphere order, the Sphaerotheriida (giant pill millipedes). Reference to continental drift can help to explain the present day distribution although it does not deal fully with all the problems (Jeekel 1974).

The glaciation of Europe had a profound effect on the fauna. Kime (1990*b*) gave a detailed account of the retreat of the deciduous forests to three main areas of Southern Europe during the last glaciation (see below). The millipedes must have retreated with the forests to these areas, or sought refuges in the form of caves and nunataks. As the ice melted, the trees advanced slowly northwards and so did the millipedes.

The three main refugia in Europe were the Iberian Peninsula, the Italian Peninsula, and the Balkans. In the Family Julidae, *Ommatoiulus* is prevalent in the Iberian Peninsula, *Cylindroiulus* and *Ophyiulus* in the Italian Peninsula, and *Pachyiulus* and *Amblyiulus* in the Balkans. The importance of the Balkans, and in particular the Carpathians, as a centre of origin for the modern diplopod fauna of Europe, has also been argued by Tabacaru (1969).

Recolonization of Norway by diplopods after the retreat of the ice was discussed by Meidell (1979). He recognized four possible routes of colonization. Golovatch (1990) considered that the impoverished fauna of the Crimea may be the result of inundation during the last Ice Age. Pleistocene glaciation in New Zealand has played a large part in shaping the diplopod fauna there (Johns 1979).

Kime (1990*b*), following a study of some elements of the British fauna, suggested that there has been some retraction of distributions since the warmest post-glacial period some five thousand years ago. This may be seen, for example, in *Polyzonium germanicum* (Fig. 3.13b) which is found presently only in the extreme south eastern corner of England (British Myriapod Group 1988).

Many species may have been distributed more widely in the past. For example, the genus *Archispirostreptus* may have once been present all across North Africa and the near East. Now, relic populations survive and have adapted to arid areas (Hoffman 1965). This phenomenon may be particularly pronounced in troglodite (cave-dwelling) species such as *Aragosoma* (= *Pyreneosoma*) *barbieri* in the Pyrenees (Mauriès 1974).

Much more recently than the last Ice Age, one agent has been responsible for fundamentally altering the distribution patterns of many millipede species. That agent is Man. The very causes of the reduced natural dispersal capabilities of the millipedes, by walking or flying, has in turn led to ease of accidental dispersion in timber, potted plants, and soil. The frequency with which 'tropical' species turn up and thrive in heated glasshouses in temperate countries is testimony to this fact.

Just to give a few examples, 59 per cent of the species recorded in Hordaland, Norway, by Meidell (1979) were considered to have been introduced there by Man. Kime (1990*b*) recorded that half the species found in Britain have been introduced into North America. In Australia,

Ommatoiulus moreleti was introduced from Portugal in about 1953 and has become a serious nuisance (Section 10.4). The reproductive strategies of some of these species are such that it is relatively easy for them to become established if the climate is suitable (Section 10.5).

2.5 Speciation

Loomis and Schmitt (1971) pointed out that millipedes have a strong tendency to speciate. Among the reasons for this they mentioned the short life cycles of the animals (this might be disputed—section 8.5). Perhaps a stronger reason is the dependence of most species on a humid microclimate which leads to easy isolation. Poor dispersal ability compounds this.

As in most other groups of invertebrates, a decrease in the number of species has been found with altitude (Simonsen 1983). Mauriès (1987) recorded a positive correlation between the number of species on each island and their areas in the French Antilles. An extensive study was carried out also on North Atlantic islands by Enghoff (1982*a*, 1983*a,b*). The island of Madeira provides spectacular examples of endemic species swarms (each swarm descended from a single introduced ancestor). In the Madeiran Archipelago, there are two such swarms from the Julida, *Acipes* with five species, and *Cylindroiulus* with 29 species.

The *Cylindroiulus* swarm on Madeira has been studied extensively. Although the original immigrant species has not been identified conclusively, it is likely to have come from the Iberian Peninsula (Read 1989). Enghoff (1982*a*) discussed the possibility of sympatric speciation in the evolution of these species but also pointed out that Madeira is a rugged island so some degree of isolation may occur.

On the Canary Islands, a similar but even more complicated situation exists with the genus *Dolichoiulus* (Enghoff, personal communication).

There are several examples of clinal variation in millipedes. One such case has been studied by Johns (1979) in New Zealand and concerns the species *Pseudoprionopeltis cinereus* (Polydesmida). This shows clines both in gonopod structure and body shape. The latter varies between that of a typical polydesmid—flat-backed with broad paranota, grey-brown colour, and living under bark—to a more cylindrical shape with reduced paranota, light brown or yellow in colour, and living in the soil. Johns (1979) suggested that the cline may have developed as the glaciers retreated after the most recent Ice Age and the animals adapted to the new climate.

Glacial retreat may have influenced the relationships between two species in the Alps studied by Pedroli-Christen (1990). Recolonization after the Ice Age probably brought the distributions of two species of *Rhymogona* (Chordeumatida) together. Where the two distributions are now touching, a hybrid zone has formed, varying in width between 500 m and several kilometres (Fig. 2.7). Within the hybrid zone, millipedes are

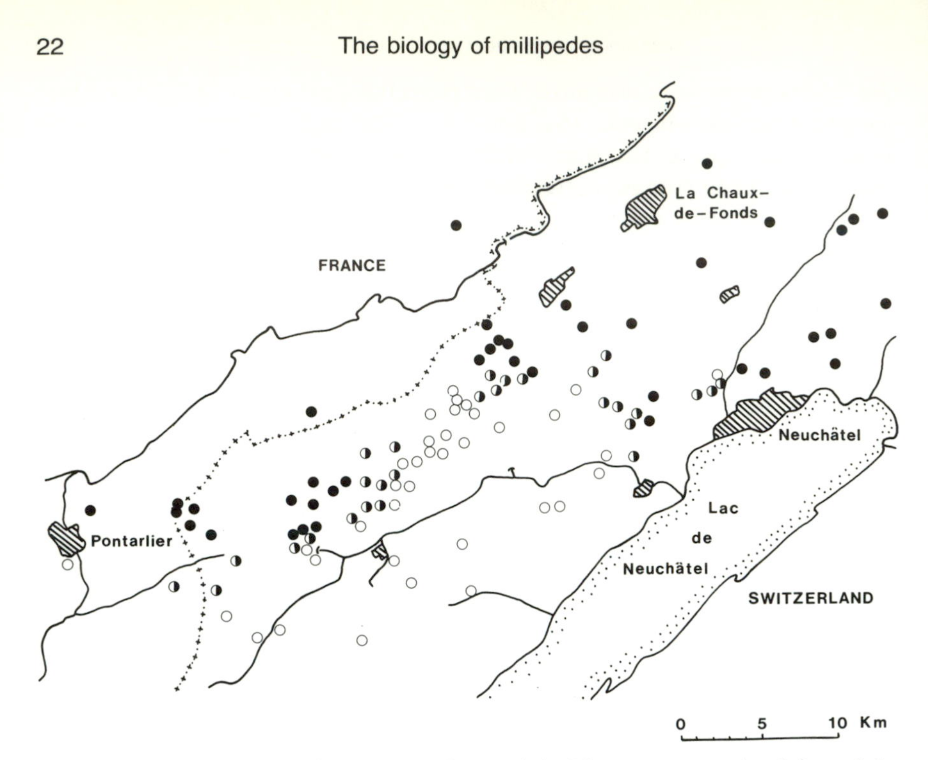

Fig. 2.7 Distribution of *Rhymogona silvatica* (○), *Rhymogona cervina* (●), and the hybrid form (◑). in the Jura. Redrawn from Pedroli-Christen (1990) by kind permission of the author and E. J. Brill, Leiden.

found that resemble one or other species, or intermediates between the two.

Where a large number of similar species are found with slightly different distributions in continental areas (due probably to parapatric or sympatric speciation), the term **mosaic complexes** has been used to describe them. Shelley (1976) made a detailed study of one complex in particular, that of the genus *Sigmoria* (Polydesmida) in the south eastern USA. In a recent publication, Shelley (1990) has reviewed the frequency of such complexes in the Diplopoda.

2.6 Mapping schemes

Several schemes for mapping the distributions of millipedes have been set up in recent years. The first national scheme to be established was in Britain. This was a joint venture between the British Myriapod Group and the Biological Records Centre of the Institute of Terrestrial Ecology at Monks Wood Experimental Station, Abbots Ripton. The scheme has developed over the years into a sophisticated system with detailed recording cards that include spaces for the recording of habitat as well as geographical information. Reports on early stages of the scheme were given by Barber and Fairhurst (1974), Fairhurst *et al.* (1978), and Fairhurst and Armitage (1979). More recently, a provisional atlas of the distribution of British millipedes has been published (British Myriapod Group 1988).

Recording schemes have been established in other countries including France (Geoffroy 1990), Hungary (Korsós 1990), and Switzerland (Pedroli-Christen, personal communication). The major task of coordinating and mapping the distributions of millipedes over the whole of Europe is being tackled by R. D. Kime. The first part of what is intended eventually to be an atlas of the distributions of all European myriapods has recently been published (Kime 1990*a*).

3

Basic anatomy, locomotion, and ecomorphology

3.1 External morphology

Millipedes consist of three main parts—the head, body, and telson.

The **head** bears the mouthparts and a number of sensory structures including the antennae, Tömösvary organs, and eyes (if present) (Fig. 3.1). The structure and function of the sense organs are covered in detail in Section 6.2. The head capsule is usually heavily calcified to facilitate burrowing between soil particles, fragments of leaf litter, or rotting wood.

The **body** is generally long and cylindrical although in some groups there are prominent lateral projections (Fig. 3.13a). The surface may be smooth

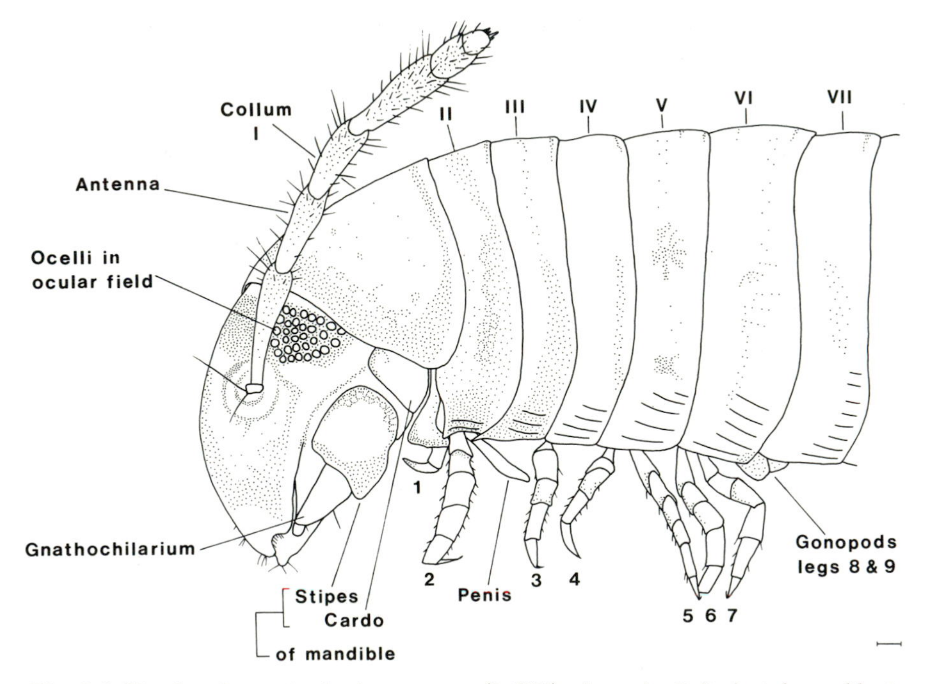

Fig. 3.1 Head and anterior body segments (I–VII) of a male *Cylindroiulus waldeni* from Madeira. Scale bar = 0.1 mm. Original drawing by H. Read.

or covered in hairlike projections, bumps, or spines. The pill millipedes are much more squat in appearance and have fewer segments. The body consists of numerous rings. Most of these bear two pairs of legs but the first three segments, consisting of the collum and rings 1, 2, and 3, share three pairs. At present, it is not completely clear as to which segments these legs 'belong' and there has been some debate in the literature on their embryological origins (see, for example, Kraus 1990). Some authors believe it is better to refer to the units of the body as 'rings' rather than segments because of the possibility that structures derived from one segment may have 'migrated' during embryological development to be associated with an adjacent segment (see Dohle 1964, 1974*b*, 1988). Since this question has yet to be resolved, the terms 'rings' and 'segments' are used interchangeably in this book.

Also present on the ventral side of the body are structures associated with sperm transfer. In all adult male millipedes except for penicillates (bristly millipedes) and Glomerida, one or both of the pairs of limbs on the seventh ring are modified to form **gonopods** (Fig. 7.6). These structures form the apparatus by which sperm is introduced into the female. Female gonopods are usually held internally but they may be extruded during copulation when they are visible behind the second pair of legs (Fig. 7.5). Exceptions to these generalizations are covered in Chapter 7. The structure of the gonopods is species-specific and is often the only reliable means of identification.

The **telson** consists of a pre-anal ring, often developed into a projection, a pair of anal plates that form a valve which opens during defaecation, and a sub-anal scale (Fig. 3.2). The shape of the projection may vary between and within species. In some cases, the sub-anal scale may also bear a projection (e.g. *Enantiulus armatus*). Between the telson and the most posterior leg-bearing ring are one or more apodous rings. Between the apodous rings and the telson lies the proliferation zone where new trunk units are initiated and develop.

The cuticle of each segment of millipedes can be considered to consist 'primitively' of a dorsal tergite and a ventral sternite, with lateral pleurites (Figs. 3.3, 3.4). However, in many groups the structure of each ring has been modified extensively. In *Polyxenus* most of the segments are double, and one diplotergite covers a pair of diplopleurites and two single sternites, one for each segment. In *Glomeris* and *Polyzonium*, tergites, pleurites, and sternites are separate, but in all other millipedes the pleurites are fused to the tergites to form the pleurotergal arch (Fig. 3.5). In julid and polydesmid millipedes, the ring is strengthened further by fusion of tergites, pleurites, and sternites into a complete structure (Fig. 3.5).

Posterior to the collum, each ring is comprised of an anterior prozonite and a posterior metazonite (Fig. 3.6). The paired defence glands, when present, usually open onto the metazonites.

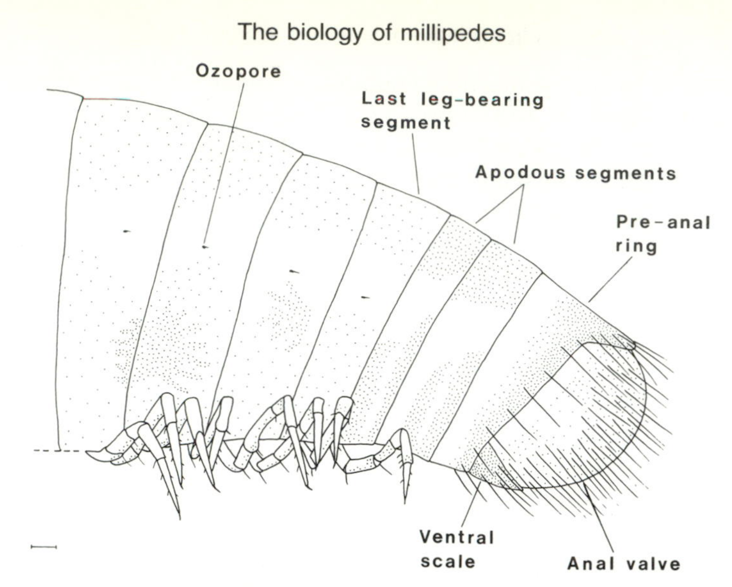

Fig. 3.2 Posterior end of a male *Cylindroiulus waldeni* from Madeira. The pre-anal ring is also known as the telson. Scale bar = 0.1 mm. Original drawing by H. Read.

Fig. 3.3 Late immobile stage of *Polydesmus angustus* (*c.* 20 mm in length) just ready for moulting. The tergites and pleurites of the future exuvia are separated (see Fig. 3.4). Reproduced from Kraus (1990) by kind permission of the author and E. J. Brill, Leiden.

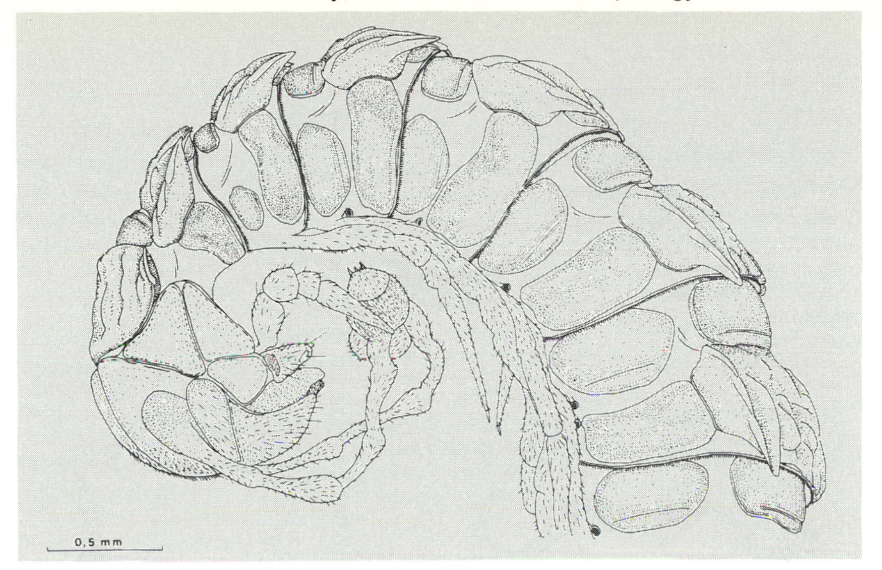

Fig. 3.4 Anterior exuvia of *Polydesmus angustus*. Lateral view of head, collum, and anterior diplosomites showing separate tergites and pleurites. Reproduced from Kraus (1990) by kind permission of the author and E. J. Brill, Leiden.

Millipede **legs** originate ventrally and this requires an S-shaped structure giving them the appearance in cross section of 'hanging down' from their legs rather than standing on them (Fig. 3.7). Each leg consists of eight sections: the coxa, trochanter, prefemur, femur, postfemur, tibia, tarsus, and claw. Some of these sections may be very short and in several species are fused together.

Millipedes have only two pairs of mouth parts, the **mandibles** and the **first maxillae** (Fig. 3.8). Each mandible consists of three parts: the cardo, stipes, and, distally, the gnathal lobe (Ishii 1988; Manton 1964). The gnathal lobe is provided with several biting and grinding devices (Enghoff 1979*a*). The structure and function of the mandibles in relation to feeding are covered in Section 4.1. Manton considered that the segmented myriapod mandible represented a whole limb but this view was criticized by Lauterbach (1972) who found evidence to suggest that the myriapod mandible may be a secondarily subdivided gnathobase. Ventrally, the mandibles are covered by the **gnathochilarium** which is formed from the first maxillae. This structure forms the 'floor' of the buccal cavity and bears a large number of sensory organs along its edge.

3.2 Cuticular structure and colouration

The cuticle of millipedes consists of three layers, a very thin **epicuticle**, the

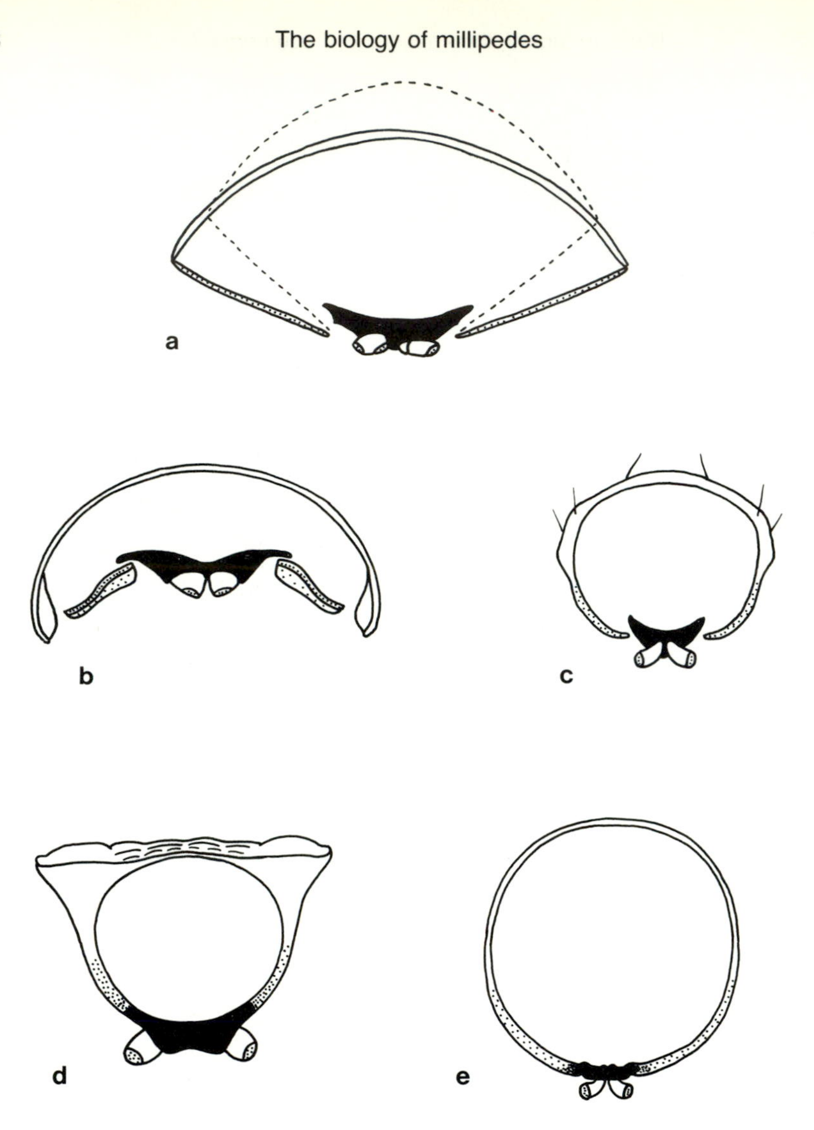

Fig. 3.5 Transverse sections showing the main types of ring structure of chilognath millipedes. (a) *Polyzonium germanicum* (Polyzoniida), (b) *Glomeris marginata* (Glomerida), (c) *Craspedosoma* (Chordeumatida), (d) *Polydesmus* (Polydesmida), (e) *Cylindroiulus* (Julida). Sternites (solid shading), pleurites (stippled) and tergites (unshaded). Scale bar = 1 mm. Redrawn from Blower (1985) by kind permission of the author, the Linnean Society of London and Academic Press.

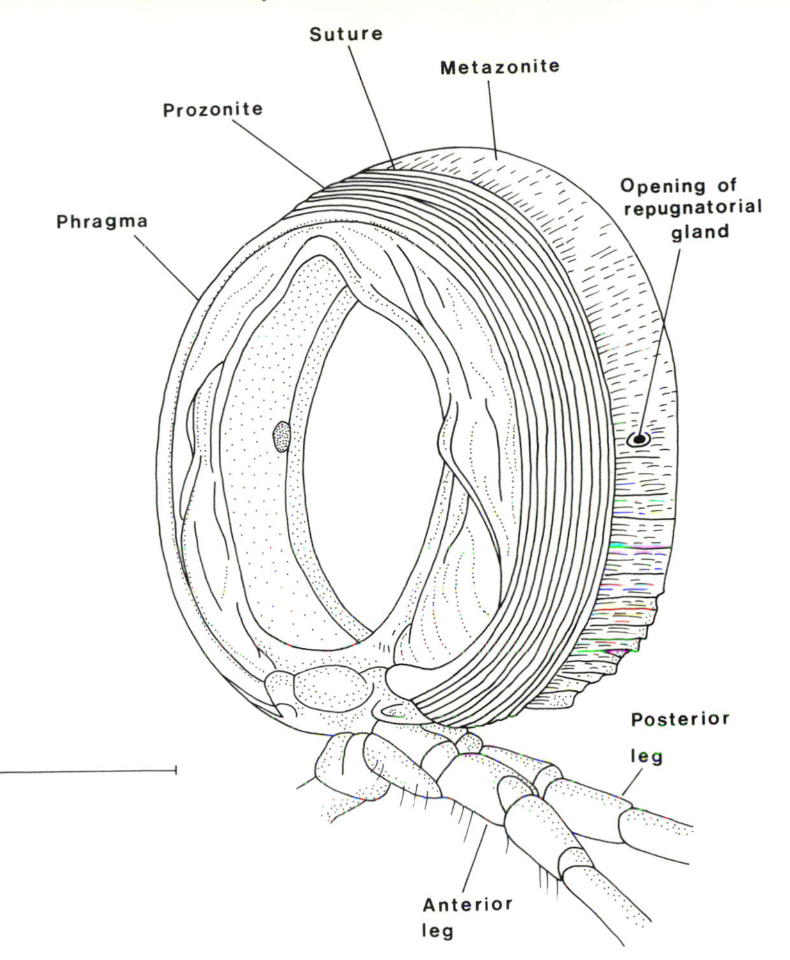

Fig. 3.6 Lateral view of basic ring structure of chilognath millipedes. Scale bar = 0.5 mm. After Demange (1981) and others.

exocuticle, and **endocuticle** (Fig. 3.9). It has been described in detail in *Tachypodoiulus niger* (Hicking 1979), *Glomeris marginata* (Ansenne *et al*. 1990), *Pachyiulus flavipes* (Carmignani and Zaccone 1977), *Spirostreptus asthenes* (Subramoniam 1974), *Orthoporus ornatus* (Walker and Crawford 1980), and *Polyzonium germanicum* (Wegensteiner 1982).

All chilognath millipedes so far examined have a calcified cuticle, although Enghoff (1990) has pointed out that this cannot be assumed to be universal until representatives of all groups have been examined in detail. The cuticle of penicillates (bristly millipedes) is not mineralized. In *Glomeris marginata*, the cuticle is mineralized with calcium and magnesium carbonates which make up 70 per cent of the dry weight of

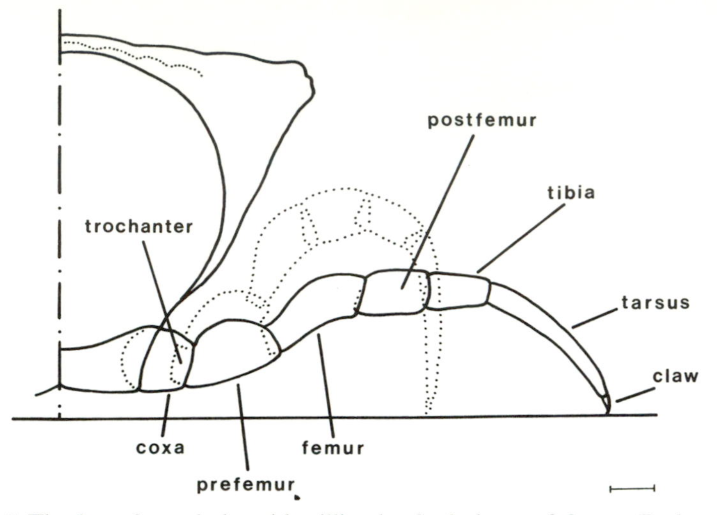

Fig. 3.7 The leg of a polydesmid millipede. Scale bar = 0.3 mm. Redrawn from Blower (1985) by kind permission of the author, the Linnean Society of London and Academic Press.

the exoskeleton. Approximately 9 per cent of the dry weight of *Julus scandinavius* and *Cylindroiulus punctatus* is calcium (Roth-Holzapfel 1990).

The need for millipedes to accumulate calcium (the only other soil arthropods to have a calcified exoskeleton are the isopods) means that they are an important component in the cycling of this essential element in terrestrial ecosystems (Cromack *et al.* 1977; Seastedt and Tate 1981). It also makes them vulnerable to accumulation of 'calcium analogues' in polluted sites, such as strontium (Gist and Crossley 1975; Krivoluckij *et al.* 1972) and fluoride (Buse 1986).

The cuticle of most species is permeable to water and this restricts their habitats to areas where the humidity is high (Edney 1977). However, some species have managed to reduce the permeability of their cuticle by incorporating a waterproofing layer of hydrocarbons (Oudejans 1972*a*; Zandee 1967). In desert millipedes, this takes the form of a thin layer of wax on the surface (Crawford 1979) which reduces water loss dramatically. For example, cuticular permeability of the desert millipede *Orthoporus ornatus* is only 7.9 $\mu g\ cm^{-2}\ h^{-1}\ mmHg^{-1}$ whereas for *Glomeris marginata* it is 200 $\mu g\ cm^{-2}\ h^{-1}\ mmHg^{-1}$ (Crawford 1972). However, care should be taken when conducting these experiments to allow for water loss via the rectum. The opposite effect, absorption of too much water across the cuticle, may be a problem in environments subject to flooding.

The epicuticle may also be capable of absorbing ant pheromones giving

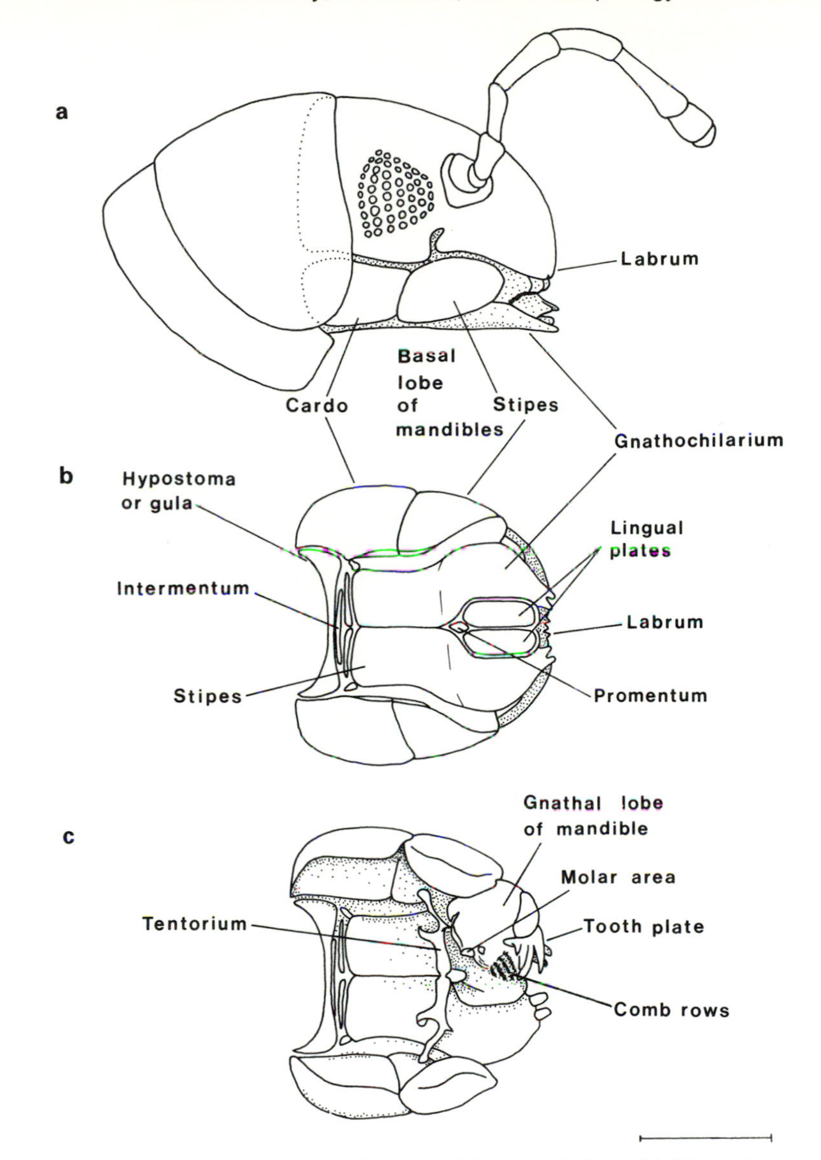

Fig. 3.8 Mouthparts of *Julus scandinavius*. (a) Lateral view. (b) Ventral view. (c) Dorsal view after removal of the dorsal part of the head capsule and clearing away the soft parts. Scale bar = 0.5 mm. Redrawn from Blower (1985) by kind permission of the author, the Linnean Society of London and Academic Press.

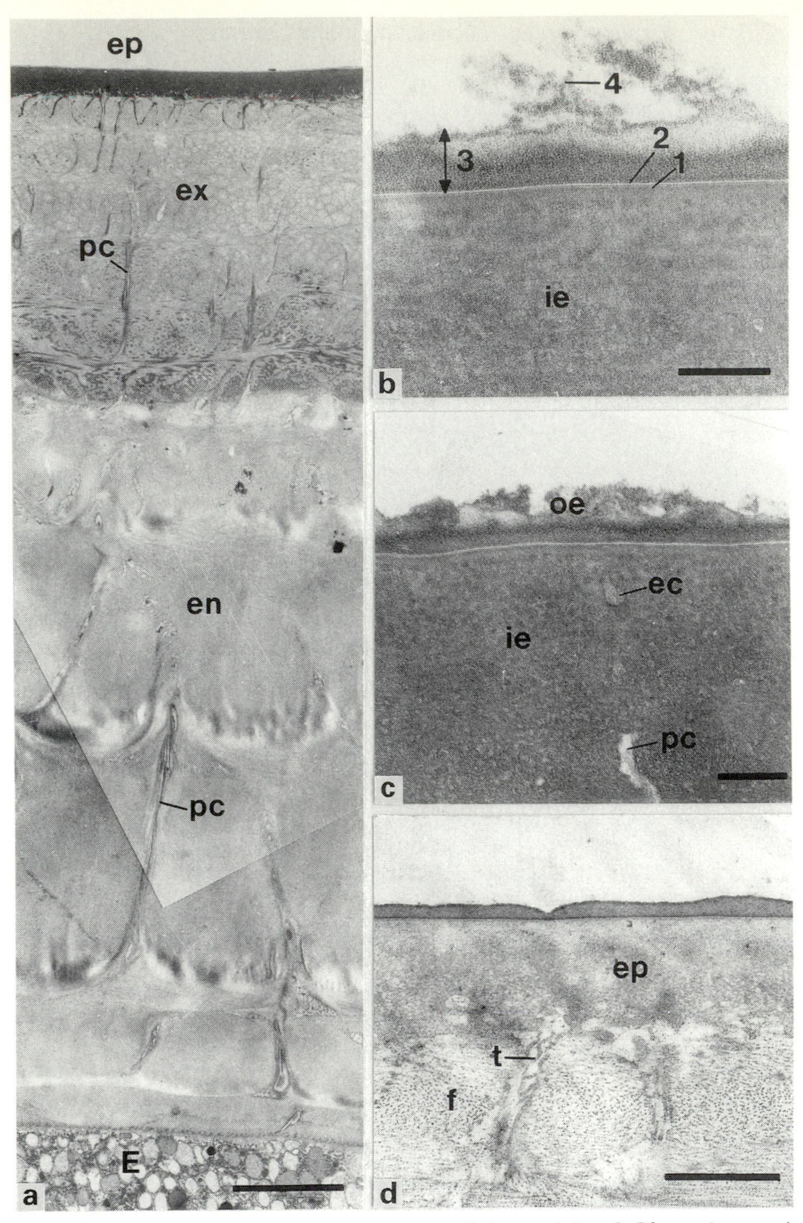

Fig. 3.9 Transmission electron micrographs of the cuticle of *Glomeris marginata*. (a) General view of tergal integument in vertical section. E, ectodermis; en, endocuticle; ep, epicuticle; ex, exocuticle; pc, pore canals. Scale bar = 5 μm. (b) Outer epicuticular layers. ie, inner epicuticle; 1, cuticulin layer; 2, lipid monolayer; 3, wax layer; 4, cement layer. Scale bar = 0.1 μm. (c) Epicuticle. ec, epicuticular canals; ie, inner epicuticle; oe, outer epicuticle; pc, pore canal. Scale bar = 0.2 μm. (d) Tubular filaments, t, in a pore canal just reaching the epicuticle, ep. Note the homogeneous distribution of the horizontal chitin-protein fibres, f, in the first procuticular lamella. Scale bar = 1 μm. Reproduced from Ansenne *et al.* (1990) by kind permission of the authors and E. J. Brill, Leiden.

some species of millipedes a 'chemical disguise' against attack by their hosts (Akre and Rettenmeyer 1968; Rettenmeyer 1962).

Most millipedes are fairly dull in appearance or show cryptic coloration. Cave-dwellers are usually white having lost all cuticular pigments. However, a few large tropical species are spectacularly coloured. One species, a spirobolid in the genus *Trachelomegalus*, was described by Lewis (1984) from Sarawak, Borneo. The collum and tail were pastel red, the legs golden yellow, and the trunk tomato red with black rings in the female, and burnt sienna with black rings in the male. When the animal coils up, a light yellow transverse dorsal stripe with a central black dot appears. Presumably this acts as warning colouration.

Many species of *Glomeris* in Europe are attractively marked with red, orange, and yellow spots. Little work has been conducted on these pigments. Needham (1968) showed that there were three main integumental pigments in *Polydesmus angustus*:

(1) coproporphyrin, which fluoresced with an intense red colour in ultraviolet light but only in dead animals (the pigment was probably bound to a protein *in vivo* which was denatured by coagulation after death);

(2) a non-fluorescent derivative soluble in carbon tetrachloride;

(3) a yellow pigment, possibly a derivative of the same metabolic pathway.

Kennedy (1978) reviewed the pigments of the Myriapoda and concluded that there were five types:

(1) quinones and the dark colours produced by tanning of cuticles

(2) ommochromes

(3) melanins

(4) flavins, and

(5) tetrapyrroles.

Some components of sensory receptors are composed of cuticle. These are described in detail in Section 6.2. Of course, millipedes have to moult their exoskeleton to grow. This aspect of their development is covered in Section 8.4.

3.3 Internal anatomy

The digestive tract of millipedes is basically a straight tube from mouth to anus (Fig. 3.10—the structure and function of the gut and associated organs are described in greater detail in Section 4.2). In some pill millipedes, the gut may be folded back on itself into a flattened S-shape to increase the length. The foregut and hindgut are lined with cuticle but the

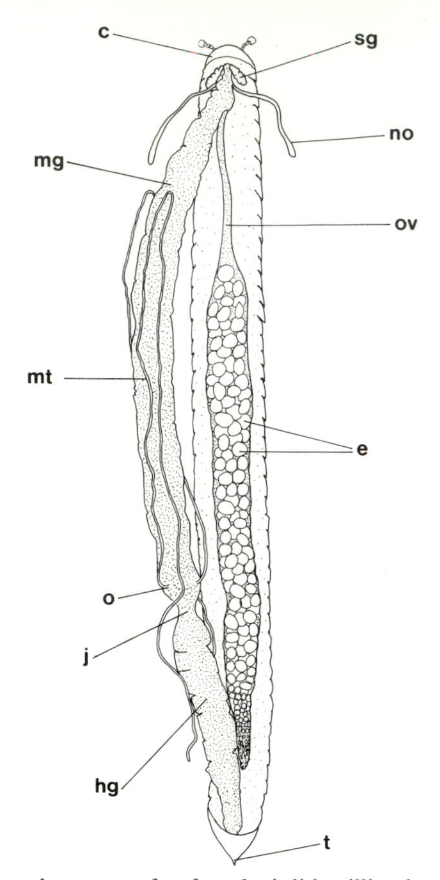

Fig. 3.10 Main internal organs of a female julid millipede of 3 cm in length. c, collum; e, eggs in ovary; hg, hindgut; j, junction of midgut and hindgut; mg, midgut; mt, Malpighian tubule; no, nephridial organ (maxillary gland); o, origin of Malpighian tubule; ov, oviduct; sg, salivary gland; t, telson. The nerve cord and fat body are not shown (after Blower 1985 and others).

midgut is not (Hubert 1977, 1979*a,b*; Nunez and Crawford 1977). Salivary glands open into the foregut and produce a secretion which lubricates the food and may contain some digestive enzymes. Other glands open into the rectum (which is eversible—Fig. 5.10) and produce secretions that are involved in the building of moulting chambers and nests. There appear to be no species of millipedes (or centipedes, pauropods, or symphylids for that matter) in which the midgut has been elaborated into a digestive gland or hepatopancreas as is found in other arthropods such as isopods.

The midgut is surrounded by a layer of cells which perform several metabolic functions. This layer has been called the 'liver' by most authors and should not be confused with the fat body, an entirely separate organ that is not associated so intimately with the midgut. The structure and function of the fat body is described in Section 5.4.10.

A pair of maxillary glands (nephridial organs) open onto the gnathochilarium. A pair of Malpighian tubules open at the midgut-hindgut junction. Both these organs are involved in osmoregulation and excretion and their structure and function are covered in Sections 5.4.6 and 5.4.9.

The reproductive organs run the length of the body and open anteriorly on the ventral side of the third body ring. In females (Fig. 3.10), the ovaries are paired and, in most millipedes, are contained within a common ovotube which bifurcates into short oviducts. These open into the vulvae behind the second pair of legs (Fig. 7.5). In males, the testes of most species consist of a pair of tubular organs. These open separately on the coxae of the second pair of legs, or through a bilobed penis just behind the legs. Further details of reproduction, including mating activity, are given in Chapter 7.

The ventral nerve cord has paired segmental ganglia (Newport 1843). In the head, the cord is swollen to form the brain, which contains a number of sites of neurosecretory activity (Fig. 6.12). These are covered in detail in Section 6.3.

3.4 Locomotion

3.4.1 Introduction

The numerous legs of the cylindrical millipedes have caught the imagination of naturalists throughout the ages. Owen (1742) stated:

> In their going, it is observable that on each side of their bodies every leg has its motion, one regularly after another, so that their legs, being numerous, form a kind of undulation and thereby communicate to the body a swifter progression than one could imagine where so many short feet are to take so many short steps that follow one another rolling on like the waves of the sea.

Later, Sinclair (1895) reported:

> It is remarkable that when the animal is in motion a sort of wave runs down the long fringe-like row of feet . . . my belief was that the feet were moved in sets of five.

Certainly, the wave-like phenomenon is one of the most endearing features of millipedes. The extensive work of Manton has given these locomotion patterns a scientific base. Her (1954) paper deals with millipedes in detail and the findings were summarized in her (1977) book.

3.4.2. Millipede legs

Male millipedes tend to have longer legs and antennae than females. This is probably related to mating activity. Longer legs enable the male to grasp the female more strongly during copulation. In general, all the legs of an individual are much the same length. However, a strange deviation occurs in *Anaulaciulus inaequipes* described by Enghoff (1986). The males of this species have much longer legs on the anterior two-thirds of the body (up to 1.78 times body diameter) than the posterior third (only 0.55 times body diameter). Again, this modification may assist grasping during copulation, although such long legs might be thought to impede the passage of the millipede through the soil.

The claw at the end of each leg is usually fairly uniform (Fig. 3.7). In julid millipedes, an accessory claw may occur also, ventral to the main claw. This accessory claw is usually slightly longer than the main claw and is setae-like in shape. The function of the accessory claw is not known but its shape does show variation between species.

Manton (1961) noted that *Dolistenus savii* has a broad claw that enables it to walk upside down and gain a purchase on smooth surfaces. Similarly, *Cylindroiulus fimbriatus* has a stout accessory claw which may help it to cling to the underside of stones (Enghoff 1983*a*).

African spirostreptids and spirobolids which are arboreal, have strong curved claws. A tree climbing julid, *Cylindroiulus lundbladi* has a much reduced accessory claw, allowing it to cling tightly to bark of trees (Enghoff 1983*a*).

3.4.3 Walking in millipedes

Walking in millipedes is rather more complicated than Sinclair's (1895) 'moving in sets of five'! When walking, each leg takes a step (Fig. 3.11). Each step consists of two stages. The first is when the claw is in contact with the ground and is moving backwards (the body of the animal moving forwards). The second is the recovery stage when the leg moves forward in the air. In millipedes, the propulsive stage (when the feet are on the ground) lasts longer than the recovery stage. The longer the propulsive stage lasts, the greater is the thrust for pushing.

Each leg pair (on the right and left sides of the body) is in phase. However, on the same side of the body, each leg is slightly out of phase with the legs immediately anterior and posterior to it. This avoids interference between adjacent legs. The phase difference is less than 0.5. Thus any given leg is less than half a cycle (of propelling and recovering) behind that in front of it. This causes the **metachronal** waves which are so characteristic of millipedes (Fig. 3.11). The number of legs between two that are in phase at any moment in time depends on the species of millipede and the speed at which it is travelling.

Speed can be increased by;

(1) taking steps more rapidly (as do polydesmids);

(2) shortening the propulsive stage in favour of the recovery phase, thereby decreasing the pushing power;

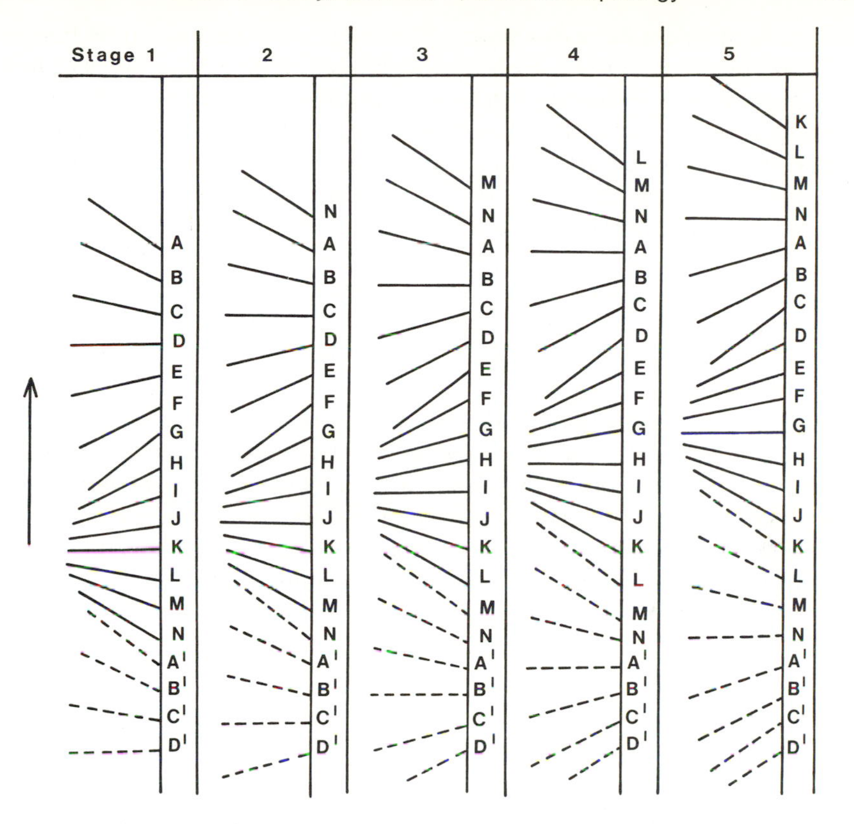

Fig. 3.11 Diagram to show the movements of the legs of the left side of a walking millipede. The legs on the right side move in phase with those on the left. The arrow shows the direction of movement of the animal. Redrawn from Demange (1981) by kind permission of the author.

(3) increasing the angle of swing that each leg moves through; or

(4) having longer legs.

3.4.4 Burrowing

Millipedes burrow by using the propulsive power of the legs to push hard. As legs of a pair are in phase, this produces an even thrust preventing undulation of the body. The more limbs that are involved, the greater is the push exerted on the head or collum which act as a pushing shield. However, large numbers of legs usually mean an extremely long body. Millipedes have reduced this potential problem by fusing the segments into diplosegments, each with two pairs of limbs, thus producing a larger thrust for a relatively shorter body.

The large forward pressure must not compress the body along the longitudinal axis, otherwise thrust will be reduced. The rings articulate by what are in effect ball-and-socket joints (Blower 1985). Like the vertebral column, the body is designed to resist shortening when all the legs contribute to the push, and yet retain the necessary flexibility to allow spiral reflexing. In other words, millipedes may still wriggle from side to side or roll into a spiral by tucking the head under the legs (Figs 3.3, 3.4, 3.12b). Pill millipedes are able to form a completely enclosed ball (Fig. 3.14a).

The internal musculature required to cope with a low gait and ability to form a spiral is extremely complex and is beyond the scope of this book. It has been described in detail by Manton (1973, 1977).

3.4.5 Jumping

A particularly unusual mode of locomotion is seen in *Diopsiulus regressus* (Evans and Blower 1973). The usual form of locomotion is as described for the borers (see Section 3.5). However when provoked, the millipedes jump approximately 2 to 3 cm, land, slide, run forward a short distance, and then jump again (Fig. 3.15). The jump is effected by humping of the body about a quarter of the way back from the head. The hump becomes a loop which is thrown upwards and forwards. The posterior half of the animal is dragged up and overtakes the anterior part. The body lands in a U-shaped position, the head is then brought round, and the animal runs forward.

3.5 Ecomorphological types

Manton was a great believer in relating structure to function. Taxonomists often describe and figure diagnostic features of their organisms without giving due thought to the reasons why a particular structure is the shape that it is. Eisenbeis and Wichard (1987), in their excellent collection of scanning electron micrographs of soil arthropods, recognized five forms of millipedes which can be regarded as 'ecomorphological' types (see Fig. 1.2). These can be viewed in the light of Manton's interpretation of functional morphology.

1. *Bulldozers or rammers* (Fig. 3.12). These are the long cylindrical 'typical' millipedes with many diplosegments. The abundance of legs enables them to push aside earth in their path like a bulldozer, using the broad head as a ram. The cuticle is hard and the rings incompressible. The body is of an even width so that the head prepares a path which the rest of the body can follow. Manton (1977) described this as the 'present day perfection of a type of burrowing'. Examples are the julids, spirobolids, and spirostreptids.

2. *Wedge types* (Fig. 3.13a). These millipedes are relatively shorter with fewer, longer legs. Each segment is expanded laterally into **paranota** or keels. The anterior end is tapered so that the animal can put its head into a

Fig. 3.12 *Ommatoiulus sabulosus* of 4 cm in length, a typical 'bulldozer' or 'rammer' ecomorphological type. (a) Head and anterior segments. (b) The same animal rolled into a spiral. Photographs by Steve Hopkin.

crevice. The legs then push upwards by straightening, causing the crevice to widen. This allows further penetration of the anterior end. This type of burrowing is useful for a lifestyle among decaying leaves. The polydesmids typify this group.

3. *Borers* (Fig. 3.13b). In contrast to the previous two groups, 'boring' millipedes have free sternites (or sternites and pleurites; Figs. 3.5, 7.1) and the joints between rings are not completely incompressible along the longitudinal axis. When the legs of one segment are on the ground and those of the next segment are off the ground, muscles pull the posterior segment onto the anterior one. As the anterior is strongly tapered, arches

Fig. 3.13 (a) Head and anterior segments of *Polydesmus angustus* of 3 cm in length, a 'wedge' ecomorphological type. (b) Head and anterior segments of *Polyzonium germanicum* of 3 cm in length, a 'borer' ecomorphological type. Photographs by Steve Hopkin.

of progressively increasing size are dragged forward, widening the crevice. Examples are the chordeumatids, and colobognath millipedes such as *Polyzonium*.

4. *Rollers* (Fig. 3.14a). The Glomerida and Sphaerotheriida are able to bulldoze quite well, but they are also able to roll into a sphere. This provides a means of defence, and reduces water loss by minimizing the surface area that is exposed to the air. The best known example of this group is *Glomeris marginata*.

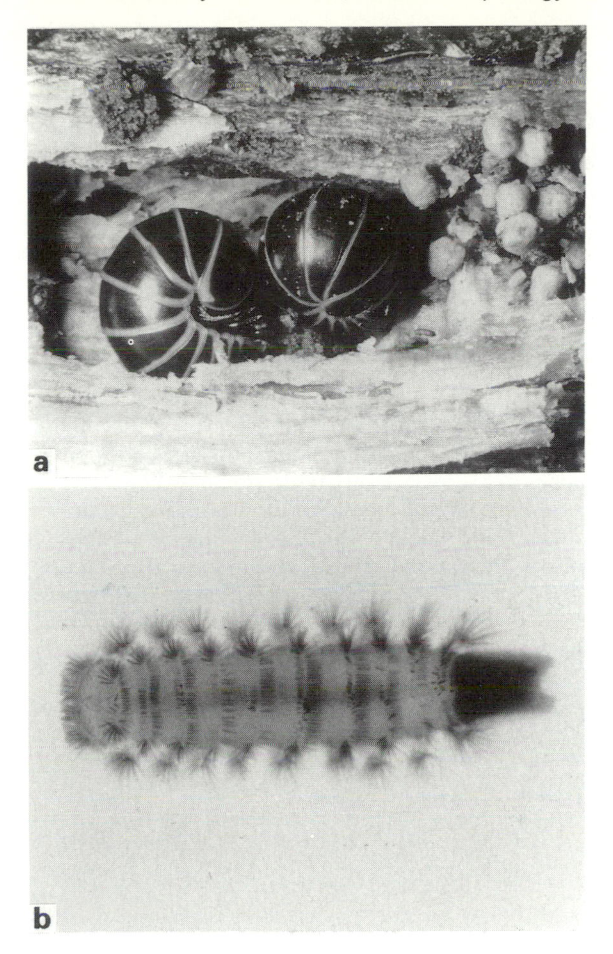

Fig. 3.14 (a) *Glomeris marginata* in rotting wood, typical 'roller' ecomorphological types. The diameter of the largest specimen when rolled up is about 7 mm. Note the large size of the faecal pellets to the right of the animals. (b) A preserved specimen of the 'bristly millipede' *Polyxenus lagurus* of 3 mm in length, a 'bark dweller' ecomorphological type (compare with Fig. 6.5). Photographs by Steve Hopkin.

5. *Bark dwellers* (Fig. 3.14b). The bristly millipedes, such as *Polyxenus lagurus*, are quite different from the 'typical' millipedes. They do not possess a strong burrowing capability. Their small size and flattened profile enable them to live in small crevices including the spaces under peeling bark. The presence of numerous serrated setae gives them a striking appearance in the scanning electron microscope (Fig. 6.5).

Of course, these categories are not strictly defined and many variations occur. For example, some julids are tapered. Enghoff (1982*b*) described an unusual species of julid with lateral tubercles and a mid-dorsal crest reminiscent of a polydesmid. In contrast, Lewis (1971*a*) described a polydesmid with reduced keels and a well-developed collum showing convergence with julids.

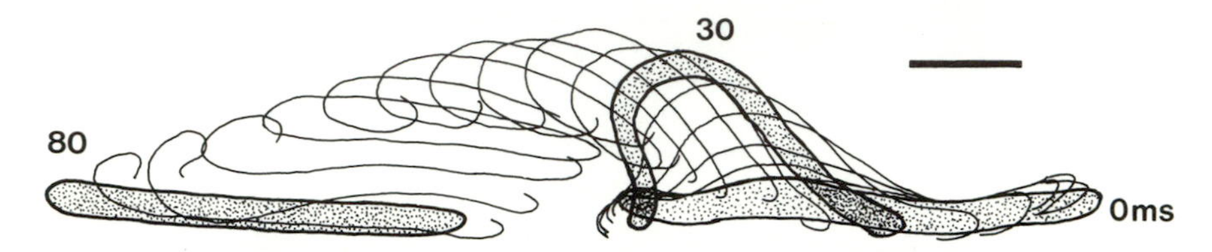

Fig. 3.15 Side view of a jump, from right to left, of the millipede *Diopsiulus regressus*. The diagram is a composite based on successive frames of a cine film taken at 2000 frames per second. Every tenth frame is shown and the figures are in milliseconds after the start of the jump. For most frames only the dorsal edge of the profile and the head and tail are shown. The initial position (0 ms) and the positions at take-off (30 ms) and landing (80 ms) are stippled. In this particular jump, the millipede rotates 180° about its longitudinal axis between take off and landing. Scale bar = 1 cm. Redrawn from Evans and Blower (1973) by kind permission of the authors and McMillan Magazines Ltd.

4

Feeding and digestion

4.1 Ingestion

In the great majority of millipedes, dead plant material is chewed by the mandibles into small pieces (Enghoff 1979*a*, 1990; Manton 1964) and passed into the lumen of the foregut where it receives secretions from the salivary glands (Nunez and Crawford 1977). The distal segment of the mandible, the gnathal lobe (Fig. 3.8), is the most complex structure of the mouthparts. The biting and crushing parts of the mandibles are based on the same general scheme but modifications exist which are taxon-specific. Exceptions to this general rule include the siphonophorids that have a long tubular beak which they may insert into plant roots (Johns 1962), and *Polyzonium* which has specialized mouthparts consisting of brush-like structures, the function of which is obscure (Kuhnelt 1976). Some platydesmids may feed on fungus using sucking mouthparts (Lewis 1984). There are even carnivorous millipedes (Hoffman and Payne 1969), and a cave-dwelling species that filter feeds (Enghoff 1985). As far as we are aware, no one has studied the internal anatomy of the these atypical species.

In millipedes that feed on dead leaves, the density of the teeth on the pectinate lamella (Figs 4.1, 4.2) determines the size of food fragments ingested (Köhler and Alberti 1990). The size of food particles which pass into the gut is correlated with digestive efficiency—the smaller the particles, the greater the proportion of food that is assimilated (Köhler *et al.* 1991).

4.2 Structure and function of the digestive tract

Digestion takes place in the midgut (Fig. 4.3). Enzymes are secreted onto the food particles by the epithelial cells and others may be derived from microorganisms in the lumen (see Section 4.6). Products of digestion are assimilated by the midgut cells and some pass through to the 'liver', a layer of cells that surrounds the midgut (Fig. 4.3).

The structure and function of the midgut has been described in *Oxidus gracilis* by Neumann (1985). He showed that the midgut was a dynamic organ, the ultrastructure of which was related to the digestive and moult cycles. The principal cell type in the midgut is the 'differentiated' cell. These cells interdigitate basally with the surrounding liver cells and contain

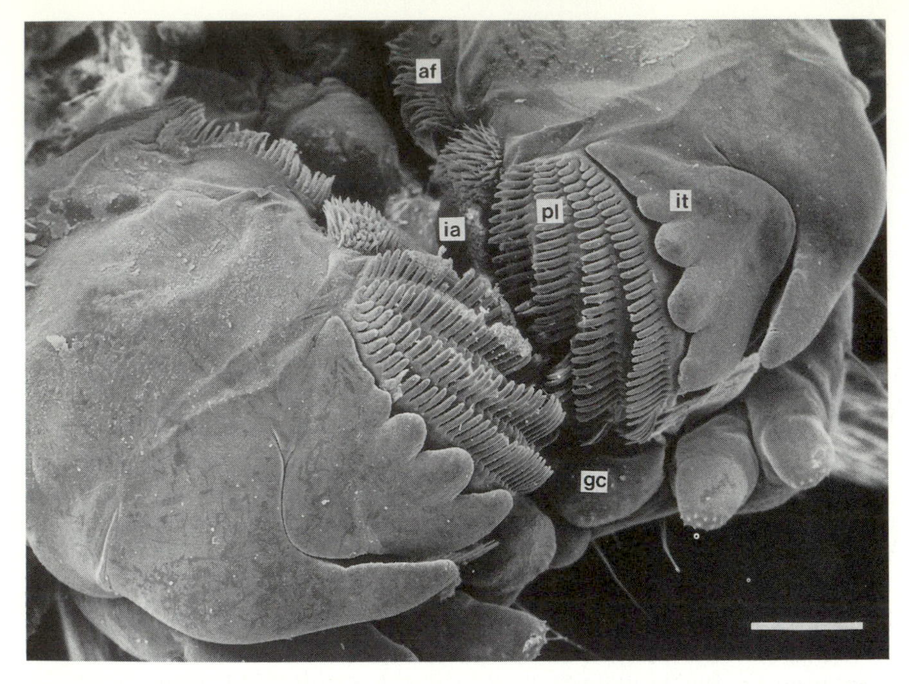

Fig. 4.1 Scanning electron micrograph of the gnathal lobes in their natural position within the oral cavity of *Julus scandinavius*. af, anterior fringe; gc, gnathochilarium; ia, intermediate area; it, internal tooth; pl, pectinate lamella. Scale bar = 100 μm. Reproduced from Köhler and Alberti (1990) by kind permission of the authors and Pergamon Press.

numerous concentrically-structured calcium phosphate granules (Figs. 4.4, 4.5). The granules may also contain other metals such as zinc (Hubert 1979*a*,*b*; Köhler and Alberti 1991). The intracellular membranes are joined at smooth septate junctions that are similar in structure to those of other arthropods (Dallai *et al*. 1990).

At moult, the entire midgut epithelium breaks down and the contents of the differentiated cells (including accumulated waste products) are excreted via the faeces. The midgut epithelium is restored after moult by the growth and division of 'regenerative' cells.

The midgut of millipedes produces a peritrophic membrane (Fig. 4.6) that surrounds food in the lumen (De Mets 1962; Hubert 1977, 1979*a,b*). The peritrophic membrane of *Glomeris marginata* has been studied recently in great detail by Martin and Kirkham (1989). A specific colloidal gold marker was developed for chitin which enabled the sites of secretion of the chitinaceous and proteinaceous components of the peritrophic membrane to be identified (Fig. 4.7).

Chitin appears first at the bases of the microvilli in synchrony along the

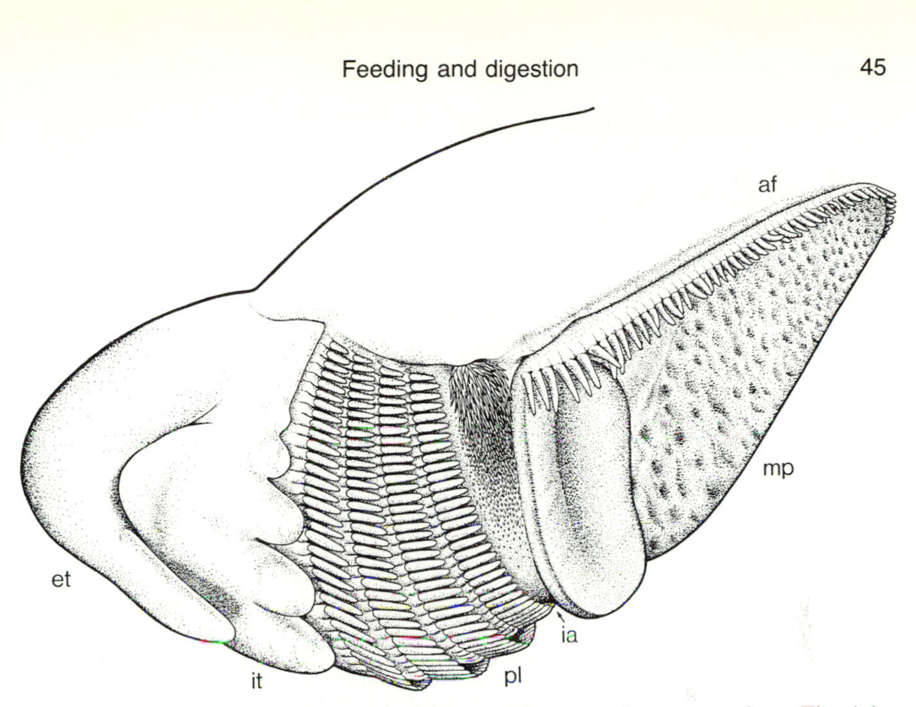

Fig. 4.2 A gnathal lobe of *Julus scandinavius* drawn to the same scale as Fig. 4.1. af, anterior fringe; et, external tooth; ia, intermediate area; it, internal tooth; mp, molar plate; pl, pectinate lamella. Reproduced from Köhler and Alberti (1990) by kind permission of the authors and Pergamon Press.

whole length of the midgut. Protein is added to the chitin from the microvilli as the peritrophic membrane moves distally along the microvilli to the lumen. The completed peritrophic membrane extends around individual items in the gut contents as well as forming a multilayered envelope. This may enhance the protective and digestive functions of the peritrophic membrane.

The midgut is surrounded by a layer of cells known as the 'liver' or 'hepatic tissue'. This probably performs functions similar to those of the midgut diverticulae or hepatopancreas in other arthropods, or may be analogous to the chloragogeneous tissue that surrounds the gut of earthworms. The cells accumulate substances which are shed eventually into the blood to be excreted elsewhere in the body. Some authors, such as Subramoniam (1972), have confused the liver with the fat body, an entirely separate organ (see Section 5.5.10).

The liver cells do not form an epithelium and are not interconnected (Seifert and Rosenberg 1977). Their basal regions ramify among the basal parts of the midgut cells (Hubert 1988). The presence of fusiform junctions suggests that open transport occurs between the two cell types. Seifert and Rosenberg (1977) showed that the liver cells of normally nourished *Oxidus gracilis* contained extensive deposits of glycogen, which were reduced rapidly if the animals were starved.

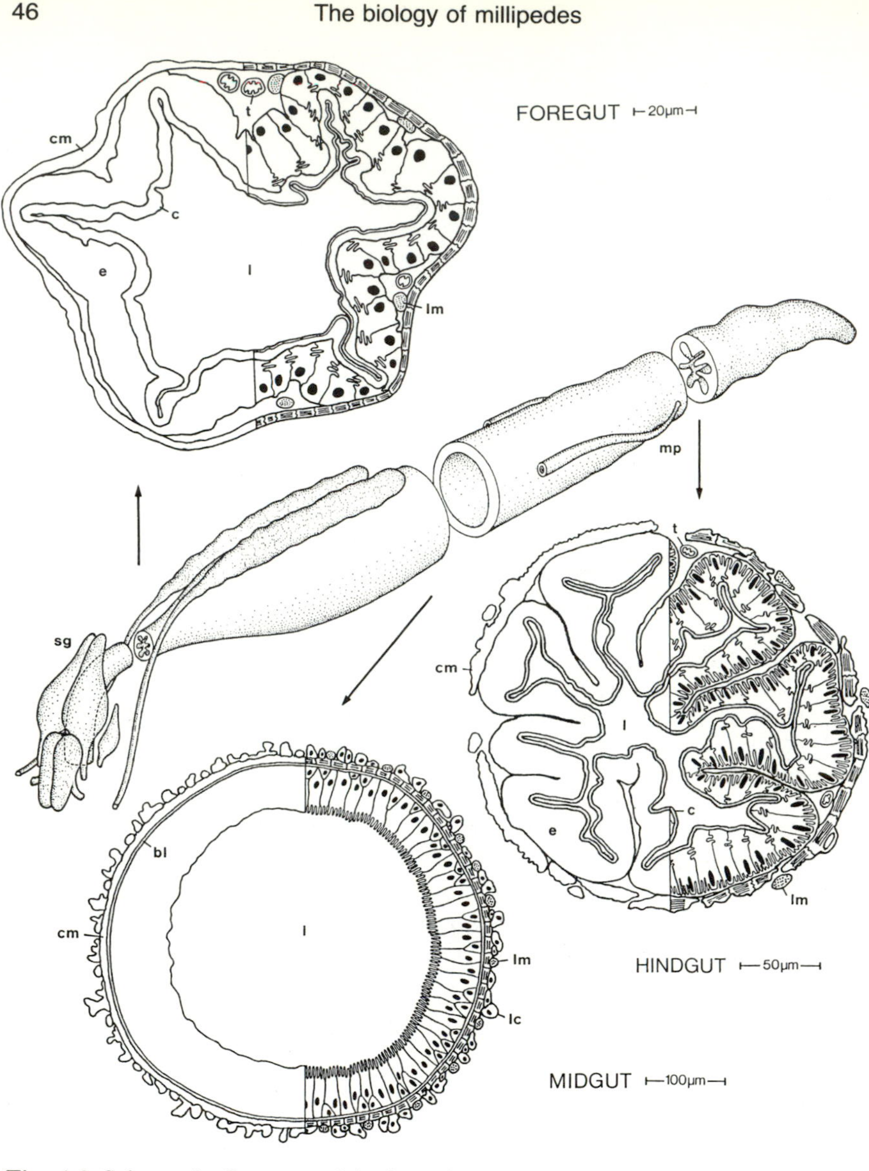

Fig. 4.3 Schematic diagrams of the intestinal tract of *Craspedosoma alemannicum* and sections of the foregut, midgut, and hindgut. bl, basal lamina; cm, circular muscle layer; c, cuticle; e, epithelium; l, lumen of intestine; lc, 'liver cell'; lm, longitudinal muscle layer; mp, malpighian tubule; sg, salivary glands; t, trachea. Reproduced from Köhler *et al.* (1991) by kind permission of the authors and the University of Innsbruck.

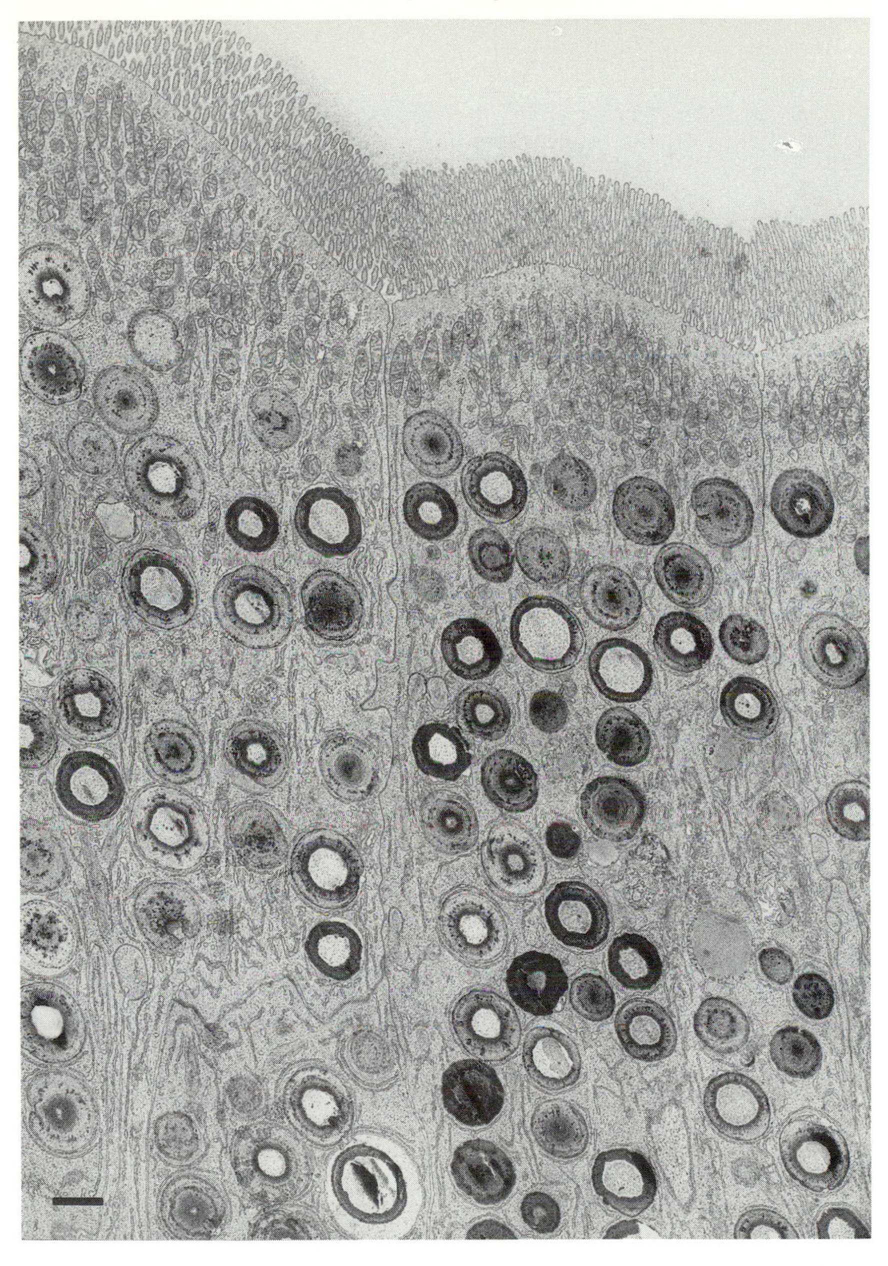

Fig. 4.4 Transmission electron micrograph of the midgut epithelium of *Mycogona germanica* (Chordeumatidae) from an uncontaminated site. Numerous concentrically structured type 'A' granules, probably containing calcium phosphate (Hopkin 1989) are present in the cells. Scale bar = 1 μm. Reproduced from Köhler *et al.* (1991) by kind permission of the authors and the University of Innsbruck.

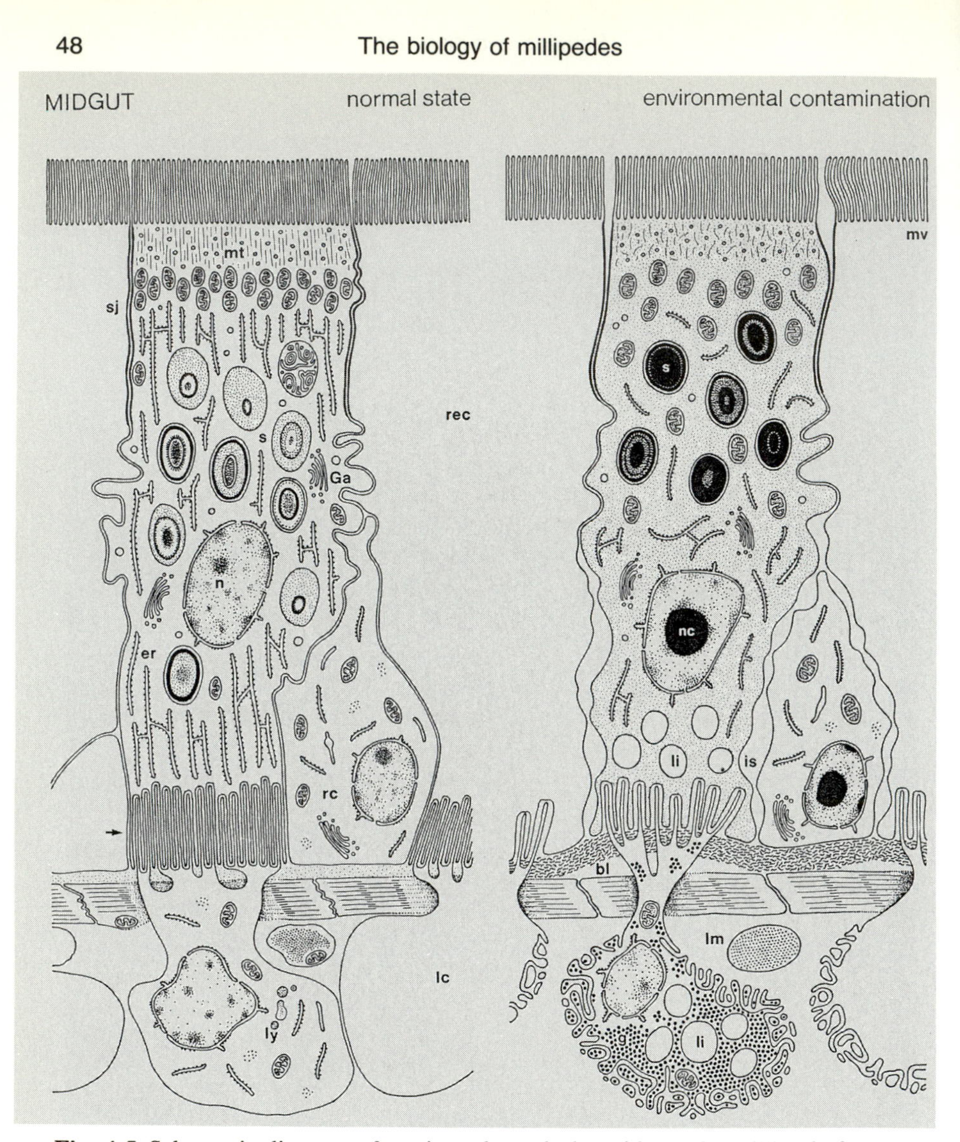

Fig. 4.5 Schematic diagram of sections through the midgut of a millipede from an uncontaminated site (left), under natural conditions in an old mining area contaminated with lead (middle), and under laboratory conditions after feeding on a lead-contaminated diet (right). The arrow indicates the zone of interaction between the resorptive cell and 'liver cell'; bl, basal lamina; cm, circular muscle layer; er,

As well as acting as sites of energy storage, the liver cells may also function as storage-excretion sites for metals and other waste substances. These are accumulated in intracellular granules (Hubert 1978*a,b*, 1979*a,b*). However, when levels of metal pollutants are very high in the diet, detoxification systems become overloaded and the cells may suffer damage (Fig. 4.5)

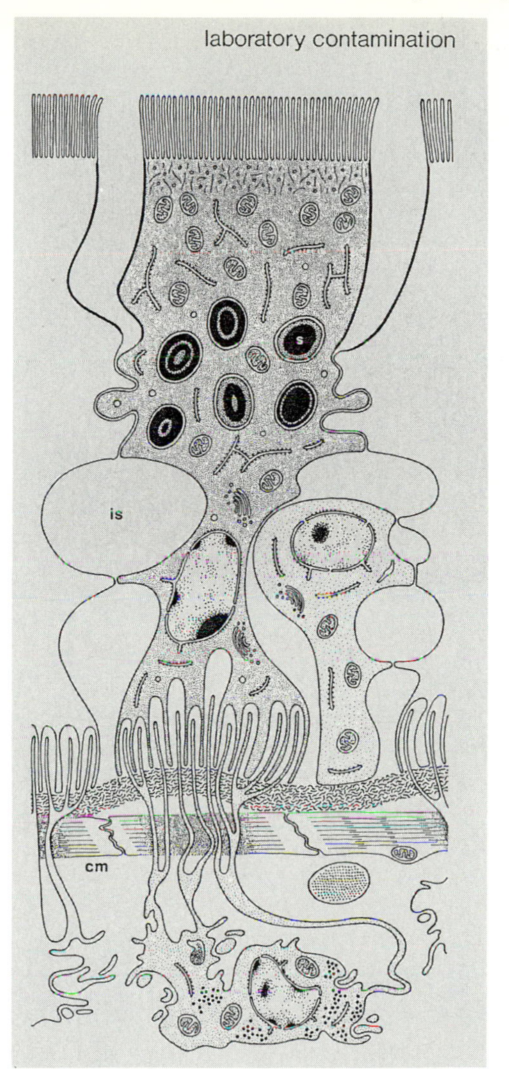

endoplasmic reticulum; Ga, Golgi apparatus; is, intercellular space; lc, 'liver cell'; li, lipid; lm, longitudinal muscle layer; ly, lysosome; mt, microtubules; mv, microvilli; n, nucleus; nc, nucleolus; rec, resorptive epithelial cell; rc, regenerative cell; s, spherite (type 'A' granule); sj, smooth septate junction. Reproduced from Köhler *et al.* (1991) by kind permission of the authors and the University of Innsbruck.

A pair of Malpighian tubules open at the midgut–hindgut junction. These are involved mainly in osmoregulation, although excretion of nitrogeneous wastes and inorganic material takes place also. The structure and function of the Malpighian tubules are described in Section 5.5.9.

The hindgut is lined with cuticle. The epicuticle of the hindgut of *Cylin-*

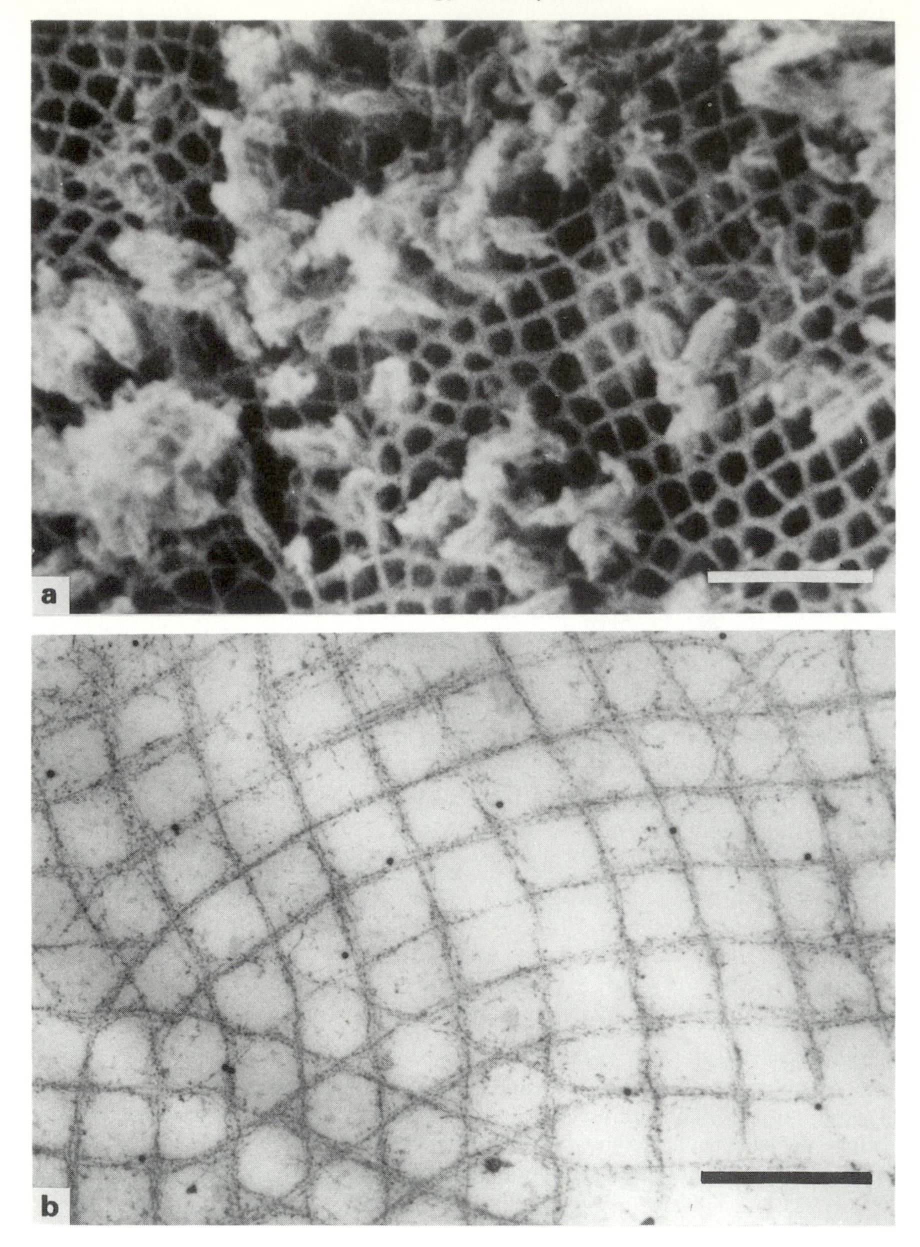
a
b

Fig. 4.6 Peritrophic membrane of *Glomeris marginata*. (a) Scanning electron micrograph of a fully formed peritrophic membrane in the midgut. The material on the surface has not been identified. Scale bar = 0.5 μm. (b) Transmission electron micrograph of a section of peritrophic membrane in the lumen adjacent to the epithelial cells in the posterior third of the midgut. The fibres are arranged in uniformly sized squares. Hexagons result where an additional row of fibres crosses the square in a third direction. The black dots are particles of colloidal gold (see Fig. 4.7). Scale bar = 0.5 μm. Reproduced from Martin and Kirkham (1989) by kind permission of the authors and Churchill Livingstone.

droiulus londinensis is pierced by tiny holes 50–100 nm across which increase the permeability (Hubert 1981).

The cuticular surface of the hindgut may be covered with spines. In *Tachypodoiulus niger*, these are of a superficial nature without anchorage in the epithelium. They probably prevent anterior movement of faecal material by gripping the remains of the peritrophic membrane (Schlüter 1980*a*). In *Polydesmus angustus*, the spines are anchored in the epithelium and are each inserted onto a longitudinal muscle which can move them. Schlüter (1980*a*) has suggested that this type of spine is able to rupture the peritrophic membrane to allow greater fluid and ion exchange with the faecal material. In *Orthoporus ornatus*, the cuticle of the hindgut is formed into numerous tiny projections, each of which has a central depression in which bacteria may reside (Crawford *et al.* 1983). The cuticular lining of the hindgut is lost with the rest of the exoskeleton at moult.

The surface of the cuticle of the hindgut exposed to the lumen may be covered in a layer of secretion derived from exocrine glands situated at the midgut–hindgut junction. These so-called 'pyloric glands' secrete their products into the longitudinal furrows of the pyloric valve and have been described in a number of species by Schlüter (1979, 1980*b*). The layer seems to influence the permeability of the hindgut and play an essential role in the reabsorption of water and other important components.

In addition to the pyloric glands, anal glands are also present in several species. In *Rhapidostreptus virgator*, Schlüter (1982, 1983) has shown that these glands are developed most highly in individuals which are about to moult. Their secretions are glue-like and are used to bind together faecal pellets and particles of soil during the construction of moulting chambers.

The hindgut has the appearance of a fluid-transporting epithelium and functions mainly in reabsorption of water from the faeces. The cells are joined at pleated septate junctions (Dallai *et al.* 1990). The hindgut cells of the Madagascan *Scaphiostreptus* sp. contain numerous plasmalemma–mitochondrial complexes which are very efficient in ion and water transport (Schlüter 1980*c*). In *Polyxenus lagurus*, the Malpighian tubules are closely attached to the hindgut and are ensheathed by an envelope consisting of several flattened cells (Schlüter and Seifert 1985*a*). This complex

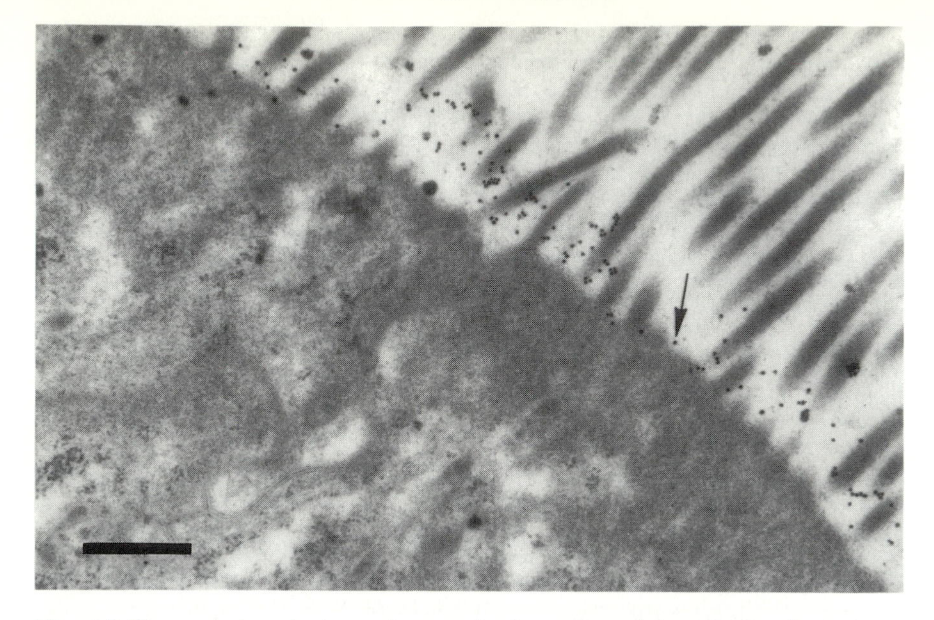

Fig. 4.7 Transmission electron micrograph of a section of the apical regions of two epithelial cells in the anterior end of the midgut of *Glomeris marginata*. Microvilli are projecting into the lumen. Particles of colloidal gold labelled with wheat germ agglutin have bound to chitin near to the sites of peritrophic membrane formation (arrow). Scale bar = 0.5 μm. Reproduced from Martin and Kirkham (1989) by kind permission of the authors and Churchill Livingstone.

(which resembles the cryptonephric system in insects) has evolved to maximize water retention (see Section 5.4.9).

The faecal pellets are formed in the rectum. The rectum can be everted and is capable of absorbing water from a moist substrate (Fig. 5.10). For example, the alpine species *Mastigona mutabilis*, when dehydrated, was able to reabsorb water via its everted rectum at a rate of 79 per cent of its body weight per hour (Meyer and Eisenbeis 1985).

4.3 Diet

The vast majority of millipedes eat dead plant material and fragments of organic matter. In deserts, millipedes survive by consuming organic detritus that collects at the bases of plants (Crawford *et al.* 1987). Some millipedes will eat living plants but these usually comprise soft and easily digestible material such as bryophytes (Bailey and de Mendonca 1990), young shoots, or fine roots. In some situations, millipedes may be quite important pests of crops. *Brachydesmus superus* and the 'spotted-snake' millipede *Blaniulus guttulatus* can stunt and even kill sugar beet seedlings in the spring by aggregated feeding on young roots (Baker 1974–see also Section 10.4).

Some millipedes have a more specialized diet. *Polyxenus lagurus* grazes algae from the bark of trees (Eisenbeis and Wichard 1987). A few species are carnivorous (Hoffman and Payne 1969) and several will eat dead animal remains including snails (Srivastava and Srivastava 1967). However, the digestive system is poorly adapted for consuming a carnivorous diet in most species. Schlüter (1980*d*) showed that the gut epithelium of *Scaphiostreptus* sp. broke down within two weeks when the millipedes were forced to eat a carnivorous diet instead of their usual detrivorous fare.

It might be thought that millipedes which live in deep leaf litter are in an ideal environment of unlimited food. However, this is not necessarily the case. Millipedes have a clear preference for leaves derived from particular tree species (Kheirallah 1978). For example, Kheirallah (1979) ranked the preference of *Julus scandinavius* for five species of leaves in the order ash, sycamore, birch, beech, oak, with oak being the least-preferred. Furthermore, each leaf had to attain a certain age before it became palatable to the animals.

Such preferences have been assigned to the levels of defensive chemicals in leaves. These substances break down, or leach out of the leaves with time, rendering the leaves more palatable. However, these assumptions were questioned by Neuhauser and Hartenstein (1978) who showed that the phenolic content was not related to palatability in four species of millipedes. Such conclusions were supported by Sakwa (1974) who suggested that the concentrations of nitrogen, carbohydrates, and moisture were more important factors. Similar conclusions with regard to calcium content were reached by Lyford (1943).

Also important is the state of decay and extent of microbial colonization of food. Bacteria and fungi may increase the availability of nutrients and 'condition' leaves, making them more palatable.

Millipedes need to avoid food which might poison them. Hopkin *et al.* (1985) and Read and Martin (1990) found that millipedes avoided eating food that was contaminated heavily with zinc, cadmium, and lead. However, this does not necessarily mean that millipedes possess receptors that respond specifically to toxic levels of zinc, cadmium, and lead in their diet. A more likely explanation is that they simply avoid eating something which tastes unpleasant. Thus in toxicity experiments, or in the field, millipedes may be dying of starvation through rejection of contaminated food rather than being killed directly by toxic levels of pollution.

4.4 Coprophagy

Opinions differ as to whether it is essential for millipedes to eat their own faeces to survive. The 'external rumen' hypothesis states that some organisms must re-ingest excreta since it provides nutrients which were not available for assimilation on the initial passage of food through the gut (Price 1988). This may be because nutrients are only released into a form that can be assimilated towards the hind end of the gut where the cells can not take them up. Additionally, it may be because the activities of microorganisms in the faeces release substances that would otherwise be unavailable.

Coprophagy is certainly essential for *Apheloria montana*. If members of this species are prevented from ingesting their faeces, they die in about a month (McBrayer 1973). Millipedes with access to faeces increased body dry weight by 16 per cent in 30 days, compared to virtually no increase in weight in those denied access to their faeces. However, in other species such as the desert-dwelling *Orthoporus ornatus*, no evidence could be found that the millipedes were coprophagous (Crawford *et al*. 1987). Similarly, Bignell (1989) never saw *Glomeris marginata* eating its own faeces.

It is probable that in most species of millipedes, corprophagy is important when the supply of a preferred diet is limited. In such situations, grazing of microorganisms, particularly fungal hyphae from the surface of faeces, may be more important than actual ingestion of undigested leaf fragments (as is the case in terrestrial isopods—Hassall and Rushton 1982, 1985). This point should be borne in mind when experiments on this topic are being planned and results interpreted.

4.5 Digestion

Animals that consume dead plant remains rely on food of relatively poor quality for their nutrients. Millipedes deal with this food in two stages.

The first stage involves the rapid assimilation of soluble materials released from the food after mechanical breakdown by the mandibles. Nutrients pass through the peritrophic membrane in the midgut and are assimilated across the microvilli. Recent work by Kohler *et al*. (1991) has shown that assimilation is greatest in those millipedes with mouthparts that grind food into the smallest particles.

The second wave of nutrients are released following digestion of the food. Digestive enzymes are derived from the secretions of the salivary glands (Nunez and Crawford 1977), and the cells of the midgut epithelium. The enzymes pass from the midgut cells through the peritrophic membrane and may be supplemented by enzymes released from microorganisms mixed with the food in the lumen (see Section 4.6). Some desert millipedes ingest sand grains, and other more temperate species may take soil into the gut, possibly to strip off adhering organic material.

In many millipedes, enzymes have been detected that are capable of digesting lipids, proteins, and simple carbohydrates (Kaplan and Hartenstein 1978; Marcuzzi and Turchetto-Lafisca 1977; Neuhauser and Hartenstein 1976; Neuhauser *et al.* 1978; Nielson 1962; Nunez and Crawford 1976). However, there is some controversy as to whether millipedes are able to digest the more refractory components of leaves.

For example, Neuhauser *et al.* (1978) showed that *Oxidus gracilis* was unable to degrade [^{14}C]-, methoxy [^{14}C]-, and side chain [^{14}C]-lignin to $^{14}CO_2$ over 10 days. This provided strong evidence that the millipedes were unable to degrade lignin. In contrast, *Pseudopolydesmus serratus* was able to degrade components of ligneous compounds, even when these were injected into the blood, by-passing the possible contribution of gut microbes to digestion (Neuhauser and Hartenstein 1976).

Similarly, there is conflicting evidence regarding cellulose breakdown. Jocteur Monrozier and Robin (1988) considered that *Glomeris marginata* was unable to digest cellulose. However, Beck and Friebe (1981) were able to demonstrate that gut extracts from *Polydesmus angustus* hydrolysed cellulose, hemicelluloses, and pectin. Cellulases in these gut extracts were probably derived from microorganisms (see Section 4.6). In desert millipedes, cellulytic activity in the gut is almost certainly due to bacteria (Taylor 1982*a*).

Cleavage of aromatic rings during digestion of food by millipedes is due also to the activities of microorganisms in the lumen rather than enzymes derived from the midgut epithelium or salivary glands (Kaplan and Hartenstein 1978). Care should be taken when conducting such experiments since cyanide in the defensive secretions of some species may interfere with cellulase activity if it leaks out (Hartenstein 1982).

The efficacy of digestive processes can be increased by physical and chemical means. Crawford *et al.* (1987) maintained that spirostreptid millipedes in deserts bask in the sun to increase their body temperature and improve the efficiency of digestion. It is otherwise difficult to explain this activity as it renders the millipedes more vulnerable to predation and water loss.

The pH of the contents of the lumen of the midgut of millipedes is rarely more than one unit either side of neutrality (pH 6 to 8) and the gut environment is aerobic (Bignell 1984*a*). Mean redox potentials in the lumen of the digestive tract of *Glomeris marginata* are + 232 mV in the midgut and + 204 mV in the hindgut.

4.6 Microorganisms

All authors agree that microorganisms play a crucial role in digestive processes in millipedes (for a review see Anderson and Ineson 1984). Indeed Crawford (1988) has suggested that the behavioural ecology of desert millipedes (in terms of temperature preferences etc.) may depend to a large extent on the optimum conditions for digestive activities of their gut microorganisms. Although it is beyond question that microorganisms are of vital importance in digestion (particularly with regard to cellulose digestion), there is no evidence that millipedes possess a permanent symbiotic microflora similar to that of termites.

The digestive efficiency of millipedes is reduced if antibiotics are incorporated into the food (Taylor 1982*b*). The microorganisms they contain seem to be those which occur at low levels in the wild. These proliferate in the ideal conditions of the gut lumen. However, Citernesi *et al*. (1977) have stressed caution—just because a microorganism is found in the gut does not necessarily mean that it performs any useful role.

Price (1988), in a fascinating review of the evolution of interactions between microorganisms and invertebrates, has pointed out that for most of evolutionary time the only substrates available for colonization were soil, water, and sediments. As larger organisms evolved, microorganisms were utilized as mutualists, providing complex biosynthetic pathways which facilitated extensive adaptive radiation. Price has summed this up simply as 'ecology recapitulates phylogeny'. Thus, present food webs reflect ancient associations. The oldest taxa form the base of the terrestrial food web, feeding on organic debris, with progressively more macroscopic forms at higher trophic levels.

Millipedes, being a relatively ancient taxon, can be fitted easily into this concept. They feed on dead plant material and 'use' microorganisms in the lumen of the gut to assist their digestion of food. Indeed the gut of millipedes provides an ideal environment for microorganisms. The lumen is protected from the vagaries of the outside environment, is permanently moist, is buffered to fairly constant pH and redox potential, and its contents are mixed thoroughly by muscular action (Bignell 1984*b*). Consequently, a wide range of microorganisms can be found. Baleux and Vivares (1974) detected species of *Klebsiella, Sarcina, Bacillus*, and *Corynebacterium* in the gut of *Ommatoiulus sabulosus*.

Leaves are made more palatable to millipedes by the activities of fungi and bacteria. Anderson and Bignell (1982) reported that *Glomeris marginata* refused to eat leaves from which the microorganisms had been washed. The microorganisms degrade polyphenols and tannins, and increase the availability of easily assimilable nutrients (Sakwa 1974). Large numbers of microorganisms are 'stripped' from leaf surfaces as the food fragments pass through the gut. These microorganisms form an important source of nutrients for millipedes as the experiments of Bignell (1989) have shown (Table 4.1).

Although some types of bacteria are digested, others may proliferate during the passage of food through the gut. In *Glomeris marginata*, there is a 100–fold increase in total numbers between the food and faeces (Table 4.2; Anderson and Bignell 1980; Anderson and Ineson 1983; Ineson and Anderson 1985; Hanlon 1981*a,b*).

Fungal hyphae also form an important food source for millipedes and

Table 4.1 Distribution of radioactivity following ingestion by *Glomeris marginata* of beech leaf litter discs singly coated with ^{14}C-leaf fibre or ^{14}C-labelled microbial tissues (mean ± standard deviation, $n = 5$ animals per determination). Analyses took place 24 hours after presentation of the discs to the animals. Reproduced from Bignell (1989) by kind permission of the author and Pergamon Press.

Location of label	^{14}C-substrate for labelling	Dis min^{-1} in haemolymph and tissues	Dis min^{-1} in gut and faeces	Dis min^{-1} as Co_2	Total dis min^{-1} ingested	Assimilation efficiency (%)
Leaf fibre	CO_2	17369 ± 4956	78945 ± 24875	361 ± 214	96676 ± 27640	18.3 ± 3.3
Escherichia coli	Glucose	6554 ± 1473	2354 ± 249	683 ± 240	9594 ± 1472	74.9 ± 4.7
E. coli	Amino acid mixture	3001 ± 477	1221 ± 198	155 ± 95	4377 ± 583	72.2 ± 0.9
E. coli	Thymidine	7458 ± 3187	1084 ± 562	104 ± 98	8645 ± 3144	86.2 ± 9.0
Er. herbicola	Amino acid mixture	27004 ± 15832	902 ± 141	325 ± 270	28231 ± 15525	95.9 ± 2.3
P. syringae	Amino acid mixture	15645 ± 4337	1113 ± 373	264 ± 64	17022 ± 4339	93.2 ± 2.2
B. subtilis	Amino acid mixture	11212 ± 2505	2389 ± 913	131 ± 133	13732 ± 2901	82.6 ± 5.0
M. hiemalis	Glucose	11455 ± 996	7849 ± 1438	5928 ± 1438	25240 ± 2433	69.2 ± 7.1
M. hiemalis	Amino acid mixture	22554 ± 4810	12206 ± 2673	11245 ± 2333	46005 ± 9711	73.5 ± 1.5

Table 4.2 Bacteria in the food, gut contents, and faeces of *Glomeris marginata*. Results are expressed as counts per gram dry weight of material. After Anderson and Ineson (1983), reproduced by kind permission of the authors and Blackwell Scientific Publications

Sample	Replicate No.	Viable bacteria (10^8 g^{-1})	Mean (10^8 g^{-1})
Litter	1	3.4	
	2	6.7	4.3
	3	2.8	
Gut	1	13.0	
	2	11.9	22.8
	3	43.5	
Faeces	1	695.5	
	2	306.9	404.1
	3	209.9	

other saprophages (Cromack *et al.* 1977). Their numbers (in terms of hyphae-forming units) decrease substantially after passage of food through the gut (Hanlon and Anderson 1980). Taylor (1982*b*) was able to isolate 76 species of fungal hyphae from the gut contents of *Orthoporus ornatus*. Actinomycetes have also been found in the gut contents and faeces of millipedes by a number of authors, but their role in digestion has not been elucidated (Chu *et al.* 1987; Dzingov *et al.* 1982; Márialigeti *et al.* 1984, 1985).

4.7 Assimilation

The importance of millipedes in the decomposition of plant remains in terrestrial ecosystems has led several authors to examine rates of ingestion and assimilation of food (for reviews of this topic see Blower 1974*b*, Köhler *et al.* 1991, and Van Der Drift 1975). However, much of this work needs to be read with care since it is not always clear what is actually being measured.

Ingestion or feeding rates are a measure of intake of food into the gut. Interpreting published figures can be difficult because there is no standard method of reporting rates of ingestion. Some workers use wet weights, others dry, and some a combination of both. Because of variability in the moisture content of food (and millipedes) we would recommend strongly that all figures are expressed on a dry weight basis. *All figures below are given on a dry weight basis unless otherwise stated*. Ingestion can then be expressed as mg of food ingested per gram body weight per 24 hours, or by giving the percentage of the body weight eaten in food per 24 hours. Thus for a millipede with a weight of 100 mg which ingests 5 mg of food in 24 hours, the figure would be 5 per cent.

Using this latter measure, typical rates of food ingestion vary from 5–10 per cent of body weight consumed per 24 hours (Bocock 1963; Hopkin *et al.* 1985; Reichle 1968). Higher or lower figures than this may occur depending on species, nature of the food, physiological state and size of the animal (including stage of the moult cycle), temperature, and time of year (Barlow 1957, 1960; Gere 1956; Kayed 1986*a,b*; Neuhauser and Hartenstein 1978; Van Der Drift 1951). It is important to know the mean rate of consumption if the role of millipedes in the breakdown of leaf litter is being considered.

Because millipedes spend a considerable proportion of their life moulting, annual rates of litter consumption are less than laboratory feeding experiments on intermoult animals might indicate. For example, most of the food consumed by *Ophyiulus pilosus* in development from egg to adult is eaten in about one third of the time (Blower 1974*b*). In this example, by the time the millipedes had reached maturity they had eaten about 5 times their weight at maturity in leaf litter. Similar figures have been found for other temperate species. *Glomeris marginata*, for example, consumes about ten times its own weight in leaf litter each year (Van der Drift 1975).

Once food has been ingested, digestion can begin. The substances that are released from the food by the action of digestive enzymes (and gut microorganisms) are assimilated at different rates. In *Glomeris marginata*, simple sugars are assimilated with efficiencies approaching 100 per cent and amino acids at efficiencies of up to 95 per cent (Bignell 1989). However, assimilation of the more refractory components occurs at a much lower rate. Bignell's (1989) experiments with *Glomeris marginata* fed on leaf fibres labelled with ^{14}C, showed that only 18 per cent of beech leaf fibres were assimilated (Table 4.1).

In most cases, total 'assimilation' of food by millipedes is calculated by subtracting the weight of faeces from the weight of food consumed. A wide range of values for assimilation have been reported with a probable maximum of about 30 per cent on a dry weight basis (Wooten and Crawford 1975; Crawford *et al.* 1987; Kayed 1978; McBrayer 1973). Smaller species tend to have higher assimilation efficiencies than larger ones (Köhler *et al.* 1989).

However, several errors can creep in if this approach is adopted. Moisture content of food and faeces is very variable and errors can occur if dry weights are calculated assuming a constant wet/dry weight relationship. Furthermore, food can remain in the gut for long periods after feeding has stopped. If this is not allowed for, and the duration of experiments is short, it is possible to come up with some very high (and probably unrealistic) values for 'assimilation' in excess of 50 per cent (Striganova 1971; Striganova and Prishutova 1990).

In addition, faecal analysis describes only net assimilation and does not reflect true digestive efficiency (Gist and Crossley 1975; Reichle 1969). Reichle (1969) stated that budgets of the differences between nutrient values of food consumed and faeces produced provide only an estimate of the digestive assimilation of nutrients. This approach should not be adopted where metabolic and digestive products are mixed in the faeces. Assimilation measurements are biased by the organism's turnover of materials through normal excretory and secretory processes. Faecal analysis describes only net assimilation which is equivalent to apparent digestible energy and does not represent the true digestive efficiency as it is often taken to do in the literature.

5

Metabolism, excretion, and water balance

5.1 Introduction

Metabolism is the sum total of the chemical processes that occur in living organisms, resulting in growth, production of energy, and elimination of waste material. In this chapter, five main areas of millipede metabolism are covered: Section 5.2 deals with respiration: Section 5.3 describes the composition and functions of the blood; Sections 5.4 and 5.5 cover excretion and the organs involved in excretory processes; Section 5.6 deals with osmoregulation and water balance, factors of crucial importance to the survival of millipedes; thermoregulation is discussed in Section 5.7.

5.2 Respiration

Respiration is defined as the process in living organisms of taking in oxygen from the surroundings and giving out carbon dioxide, and the chemical breakdown of complex organic substances that takes place in the cells and tissues during which energy is released and carbon dioxide produced.

Exchange of oxygen and carbon dioxide between the cells of millipedes and the atmosphere takes place via tracheae, in a similar way to insects. In general, the tracheal system opens via small holes in the cuticle called spiracles. There is a pair of spiracles on each sternite (two pairs on each diplosternite; Fig. 5.1). These are slightly anterior and lateral to the coxae of the legs (Blower 1985). In polydesmids, the openings of the spiracles are protected by a cuticular lattice (Fig. 5.2). In some species such as *Glomeris marginata*, the openings are closeable so that water loss from the moist linings of the tracheae can be controlled.

The spiracles open internally into a spacious atrium or tracheal pouch from which numerous unbranched tracheae ramify among the tissues. The tracheal pouch may serve as an apodeme (anchoring point) for the attachment of the extrinsic muscles of the coxae (Blower 1985). Because they are air-breathing, almost all millipedes must come to the surface in waterlogged soils to avoid drowning. However, in Central Amazonian forests, some species can survive complete submersion for several months since they possess a plastron (Fig. 5.3; Messner and Adis 1988). This structure holds a layer of air over the surface of the millipede and allows respiration to

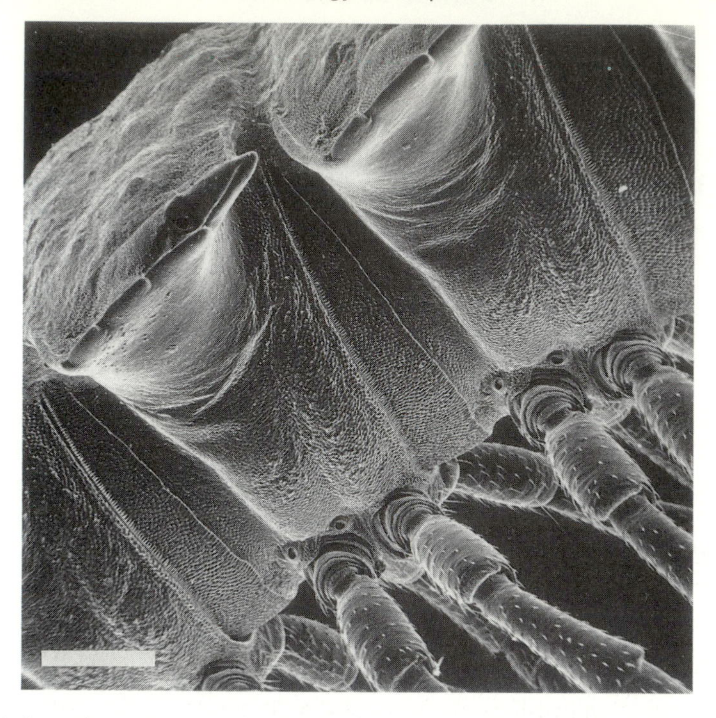

Fig. 5.1 Scanning electron micrograph (lateral view) of body segments of *Polydesmus angustus* showing spiracles near to the bases of the legs (see also Fig. 5.2). Scale bar = 250 μm. Reproduced from Eisenbeis and Wichard (1987) by kind permission of the authors and Springer-Verlag.

continue under water. In these conditions, the millipedes take up dissolved oxygen at a rate of >10 $\mu l\, mg^{-1}\, h^{-1}$.

There are a large number of publications that describe rates of oxygen uptake in millipedes. As with food ingestion rates, different authors have used different units and experimental conditions, so comparison between species is difficult. Most large millipedes are sluggish animals and have a relatively low respiration rate in comparison to other arthropods (Table 5.1), unless of course they are disturbed or forced to run rapidly (Reichle 1968).

Simply measuring oxygen consumption rates is of little value unless they are put into context by comparison with other physiological or ecological parameters. In one such example, Wooten and Crawford (1974) made monthly measurements of the respiration rate of the desert millipede *Orthoporus ornatus* for a year, and compared these with behaviour in the field (Fig. 5.4). They showed that changes in use of metabolic reserves followed changes in ambient temperatures in the field and peaked in July when the animals were on the soil surface. They estimated annual respiratory metabolism to be 5535 $J\, g^{-1}\, y^{-1}$.

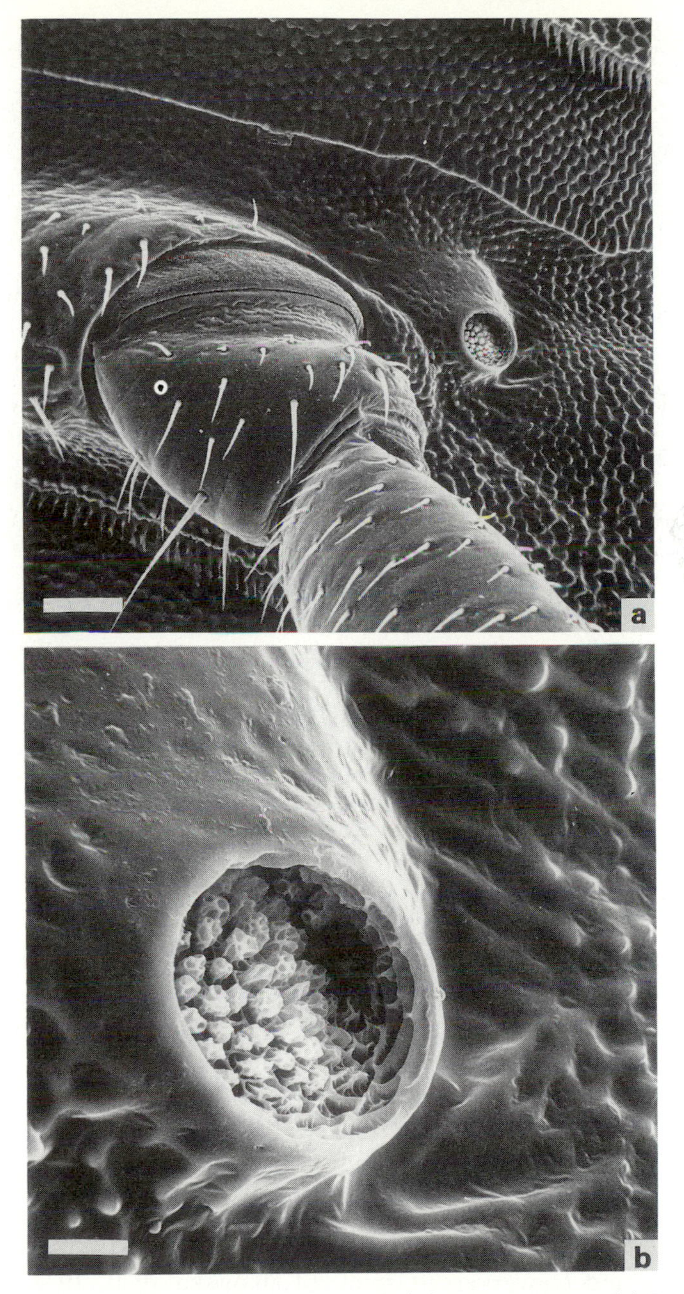

Fig. 5.2 Scanning electron micrographs of *Polydesmus angustus*. (a) Base of leg showing coxa, trochanter, and spiracle opening. Scale bar = 50 μm. (b) Higher magnification view of the spiracle shown in (a). The opening is protected from the intrusion of extraneous particles by a cuticular lattice or network. Scale bar = 10 μm. Reproduced from Eisenbeis and Wichard (1987) by kind permission of the authors and Springer-Verlag.

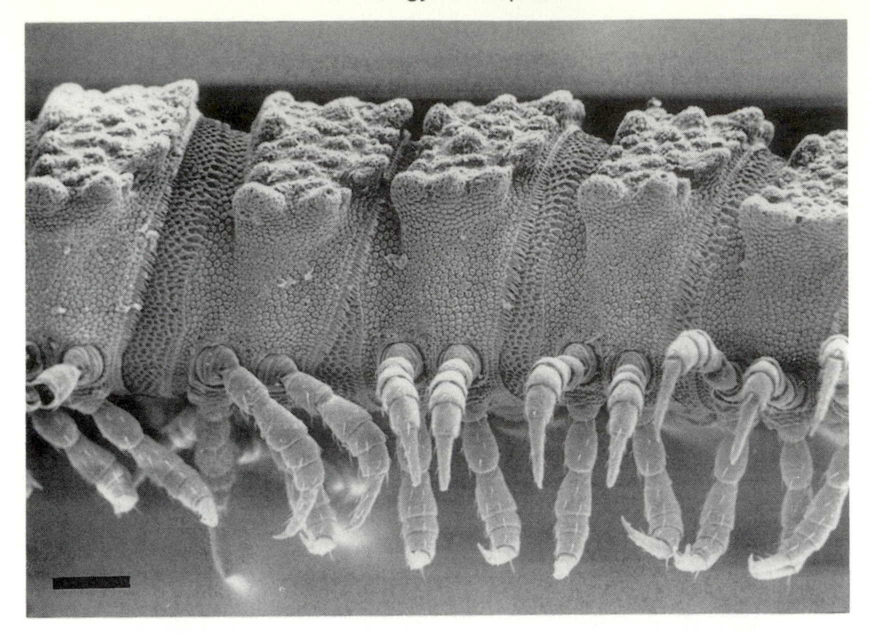

Fig. 5.3 Scanning electron micrograph (lateral view) of body segments of *Muyudesmus obliteratus* (Pyrgodesmidae). The complicated structures on the surface of the cuticle maintain a layer of air over parts of the body when the animal is submerged under water (= incomplete plastron). Scale bar = 100 μm. Reproduced from Messner and Adis (1988) by kind permission of the authors and Gustav Fischer Verlag.

Penteado (1987) examined the respiratory responses of the tropical spirostreptid *Plusioporus setiger* to decreased oxygen tension. He showed that the species regulated respiration down to 35.4 mm Hg 0_2 when suddenly exposed to a decreased oxygen tension, or 17.7 mm Hg 0_2 when the decrease was stepwise. Oxygen dependence indices (*K1*/*K2*) were relatively low, also showing regulation, but no relation to size (weight) was recorded. After hypoxia, *P. setiger* showed a typical pattern of 'under repayment' on the return to normoxia. Similar results were found in *Pseudoannolene tricolor* (Penteado and Hebling-Beraldo 1991).

More details on respiration in millipedes can be found in the recent review by Penteado *et al.* (1991), the series of papers by Gromysz-Kalkowska and co-workers (Gromysz-Kalkowska 1970, 1974, 1976*a,b*, 1979, 1980; Gromysz-Kalkowska and Stojalowska 1983; Gromysz-Kalkowska and Tracz 1983; Gromysz-Kalkowska *et al.* 1986), and those of Dwarakanath (1971) and Stamou and Iatrou (1990).

Table 5.1 Resting metabolism of 13 species of forest floor arthropods at 20°C. Reproduced from Reichle (1968) by kind permission of the author.

Class	Species	No individuals	$\bar{x}$ live wt. mg	µl 0_2 h^{-1} (per individual)	µl 0_2 g^{-} h^{-1}
Isopoda					
	Armadillidium vulgare	30	65.5	9.58	146.2
	Armadillidium nasatum	11	30.0	3.87	128.9
	Cylisticus convexus	29	34.8	7.13	205.0
	Metoponorthus pruinosis	18	18.4	6.14	333.9
Orthoptera					
	Acheta domesticus	4	276.0	55.61	201.5
	Ceuthophilus gracilipes	5	259.1	68.14	263.0
	Parcoblatta sp.	6	73.0	11.53	158.0
Diplopoda					
	Dixidesmus (=*Pseudopolydesmus*) *erasus*	20	77.6	10.85	139.8
	Cambala annulata	9	571.6	33.72	59.0
	Ptyoiulus impressus	5	431.5	24.34	56.4
Coleoptera					
	Popilius disjunctus	4	1630.5	202.02	123.9
	Evarthrus sodalis	8	164.1	23.40	142.6
	Sphaeroderus stenostomus	5	169.6	32.85	193.7

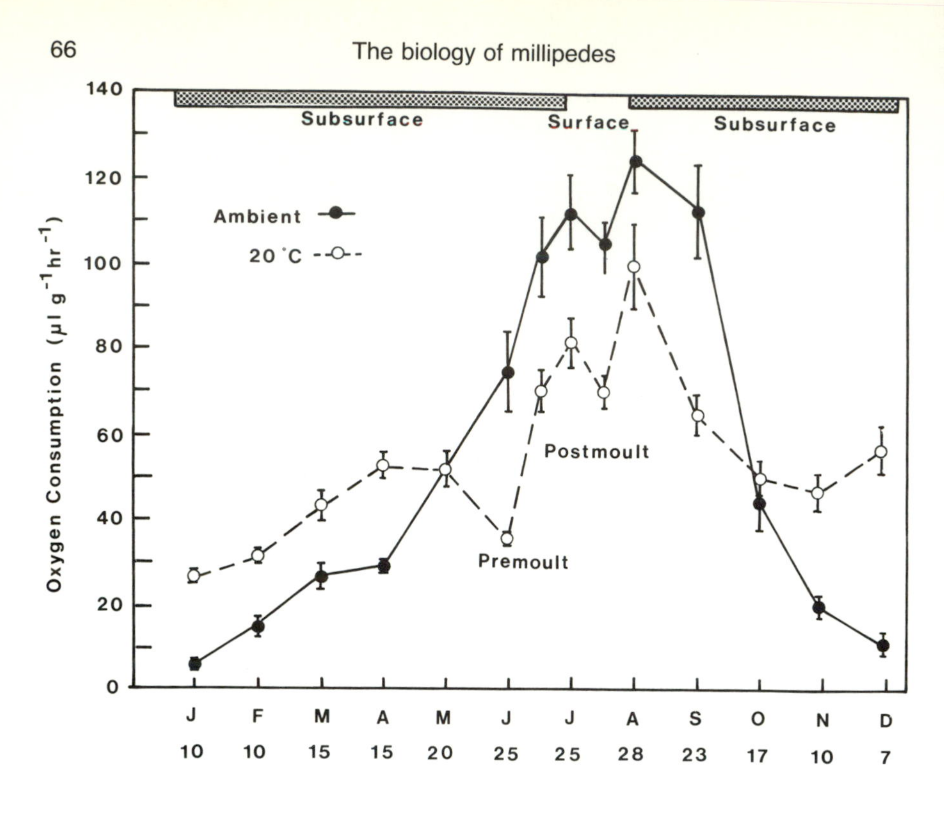

Fig. 5.4 Seasonal differences in mean respiratory rate per unit live weight of *Orthoporus ornatus* collected from nests of the harvester ant *Novomessor cockerelli* (except for July surface specimens) at Jornada Validation Site. Oxygen consumption was measured either at 20 °C (open circles) or at the average soil temperature based on IBP records and on measurements taken when millipedes were collected (filled circles). Vertical lines indicate standard error of the mean. Redrawn from Wooten and Crawford (1974) by kind permission of the authors and Springer-Verlag.

5.3 Blood

The blood of millipedes serves a number of functions. It is composed of liquid and cellular components which are circulated throughout the body by the pumping action of the dorsal vessel or heart. In *Cingalobolus bugnioni*, the heart beats at a rate of between 52 and 61 times per minute (Rajulu 1967), similar to that of a relaxed myriapodologist!

The liquid component bathes the organs of the body and transports products of metabolism to and from organs of digestion, storage, and excretion. Rajulu (1974) reported that the blood of *Cingalobolus bugnioni* contained a mean of 1.9 $mg\,ml^{-1}$ of protein, 0.86 $mg\,m1^{-1}$ of carbohydrate, and 1.5 mg ml^{-1} of lipid. The principal sugar was trehalose and phospholipids were the main lipids in the haemolymph.

The haemolymph of laboratory populations of *Triaenostreptus triodus* contained a mean of 4.9 mmol l^{-1} of amino acids (Marcus *et al.* 1987). Krishnan Nair and Prabhu (1971) detected 16 free amino acids in the haemolymph of *Jonespeltis splendidus* (=*Anoplodesmus saussurei*).

Clearly, such values will vary depending on the physiological state of the animal and whether it has just fed. The typeof food may also be important. In *Orthomorpha* (=*Oxidus*) *gracilis*, millipedes fed on their preferred diet of *Ficus sycamore* had higher concentrations of haemolymph proteins than those fed on *Hedera helix* (Kheirallah and Shabana 1975). In a detailed study on *Orthoporus ornatus* Pugach and Crawford (1978) described the seasonal changes in the amino acid, protein, and inorganic ion composition of the haemolymph (Tables 5.2, 5.3, 5.4).

The blood also transports nitrogeneous wastes in a form that can be tolerated by the millipede, and which can be excreted. Excretion is achieved via organs that discharge to the lumen of the gut (Malpighian tubules), or directly to the outside (nephridia), or to organs of long-term storage. The latter includes the fat body where nitrogeneous waste products are stored as large deposits of potassium urate (Hubert 1979*a,b*).

The concentrations of some components of the blood are almost certainly under neuroendocrine control. Satyam and Ramamurthi (1978) managed to increase the glucose concentration of the blood of *Spirostreptus asthenes*

Table 5.2 Seasonal changes in mean haemolymph concentrations of amino acids and protein in *Orthoporus ornatus*. Reproduced from Pugach and Crawford (1978) by kind permission of the authors.

			Amino acid		Protein
Collection date	Collection site soil temperature (± 3°C)*	*N*	Milligrams per 100 ml ± SE†	*N*	Milligrams per 100 ml ± SE†
6 Oct. 1974	20	11	34.79 ± 3.83 *ae*	15	11.63 ± 1.52 *a*
12 Dec. 1974	10	6	22.53 ± 1.41 *b*	6	19.98 ± 4.73 *ad*
14 Jan. 1976	10	18	32.10 ± 2.18 *e*		—
24 Mar. 1975	13	8	38.05 ± 6.21 *ace*	7	37.56 ± 4.75 *bc*
8 Apr. 1975	15	7	51.26 ± 3.89 *cd*	7	37.30 ± 6.33 *c*
25 May 1975	20	7	46.44 ± 6.59 *ad*	8	24.96 ± 4.67 *cd*
22 June 1975	25	15	61.73 ± 5.08 *d*	16	14.13 ± 2.93 *ad*
14 Aug. 1975	—	28	49.69 ± 4.68 *ad*	27	19.50 ± 2.46 *d*

* Range of soil temperatures at collection sites. As specimens were collected from soil surface, shrubs, and rock crevices in August, soil temperatures at time of that collection are meaningless and not given.
† Concentrations given are means of *N* individual values within each season. Means followed by at least one common letter are not significantly different from each other at the $P < 0.05$ level.

Table 5.3 Seasonal composition of inorganic ions in the haemolymph of *Orthoporus ornatus*. Reproduced from Pugach and Crawford (1978) by kind permission of the authors.

Collection date and soil temperature (± 3°C)	Mean concentration, mM ± SE†							Summed means of ions
	Cations					Anions		
	N	Na	K	Ca	Mg	*N*	Cl	
24 Dec. 1974 10°C	7	90.11 *ab* ±7.71	4.51 *a* ±0.41	11.37 *a* ±0.66	8.90 *ab* ±0.67	7	91.30 ±5.47	206.19
14 Jan. 1976 10°C	18	102.32 *c* ±2.18	5.74 *b* ±0.22	17.52 *c* ±0.59	11.79 *cd* ±0.57	15	69.80 ±1.82	207.17
28 Mar. 1976 16°C	20	99.03 *acd* ±1.73	5.40 *ab* ±0.23	13.68 *d* ±0.34	9.80 *bde* ±0.77	5	95.07 ±10.42	222.98
8 Apr. 1975 15°C						4	77.63 ±5.77	
25 May 1975 20°C	8	90.69 *a* ±4.48	6.01 *bc* ±0.41	12.30 *ad* ±0.39	7.08 *a* ±0.57	8	65.08 ±3.33	181.16
22 June 1975 25°C	16	76.94 *b* ±2.95	5.31 *ab* ±0.36	11.01 *a* ±0.85	11.56 *bd* ±1.32	8	65.63 ±4.38	170.45
14 Aug. 1975*	29	94.28 *a* ±1.66	6.52 *c* ±0.17	9.28 *b* ±0.19	8.04 *ac* ±0.40	23	59.69 ±2.05	177.81

* See footnote in Table 5.2.

† Concentrations given are means of *N* individual values within each season. Concentrations followed by at least one common letter within a column are not significantly different from each other at least at the $P < 0.05$ level.

Table 5.4 Seasonal contribution of inorganic ions to haemolymph osmolality in *Orthoporus ornatus*. Reproduced from Pugach and Crawford (1978) by kind permission of the authors.

Collection date	Osmolality, mosmol ±SE (*N*)	Contribution to osmolality, %* ± SE (*N*)					
		Na^+	K^+	Ca^{2+}	Mg^{2+}	Cl^-	Total ions
21 Dec. 1974	295.7 ±8.5 (9)	31.16 ±2.48 (7)	1.55 ±0.13	3.95 ±0.21	3.04 ±0.19	30.60 ±2.53 (7)	73.62 ±4.08 (5)
14 Jan. 1976	321.6 ±8.2 (10)	32.95 ±0.90 (10)	1.85 ±0.12	5.42 ±0.32	3.79 ±0.31	21.50 ±0.87 (8)	65.53 ±2.59 (8)
28 Mar. 1976	291.6 ±4.0 (19)	34.01 ±3.36 (19)	1.82 ±0.08	4.67 ±0.12	3.25 ±0.26	31.77 ±2.98 (5)	72.45 ±3.34 (5)
8 Apr. 1975	249.6 ±9.7 (7)					29.93 ±2.31 (4)	
25 May 1975	216.7 ±7.3 (12)	42.41 ±1.99 (8)	2.81 ±0.17	5.78 ±0.26	3.31 ±0.27	30.78 ±1.90 (8)	84.40 ±4.49 (7)
22 June 1975	211.9 ±5.9 (20)	37.49 ±1.35 (20)	2.60 ±0.19	5.40 ±0.45	5.72 ±0.72	30.34 ±1.53 (8)	80.65 ±2.70 (8)
14 Aug. 1975	204.1 ±3.9 (28)	46.88 ±0.90 (28)	3.26 ±0.09	4.61 ±0.11	3.99 ±0.18	29.55 ±1.00 (23)	87.94 ±1.60 (23)

* Percentages given are means of *N* individuals within each season. The same number of specimens was used for all cation determinations on a given date.

from 0.026 mg ml^{-1} to 0.072 mg ml^{-1} by implantation of the brain from another millipede.

The haemocytes which circulate in the blood perform a number of functions. Ravindranath (1973, 1981) recognized seven types in the haemolymph of *Thyropygus poseidon*, based on observations by light microscopy. These were:

(1) prohaemocytes;
(2) plasmatocytes;
(3) granular haemocytes;
(4) cytocytes;
(5) oenocytoids;
(6) spherulocytes; and
(7) adipohemocytes.

The relative numbers of the haemocyte types change with the moult cycle. The granulocytes contain the enzyme phenol oxidase which is involved in tanning the organic components of the cuticle (Krishnan and Ravindranath 1973). The spherulocytes are thought to transport calcium carbonate to mineralize the new soft cuticle since they are most prominent immediately after moulting (Ravindranath 1974).

The blood of millipedes has antibacterial properties which are inducible. In *Triaenostreptus triodus* (Fig. 5.5A), blood removed from control animals had no inhibitory effect on bacteria. However, blood removed from millipedes which had been inoculated 48 hours previously with *E. coli* was much more effective at destroying the same bacteria than those which had not been previously exposed (Fig. 5.5B). The active factor was represented by a single protein band (Fig. 5.5C; Van der Walt *et al.* 1990).

In *Chicobolus* sp. and *Rhapidostreptus virgator*, the antibacterial substances were unstable when heated but were resistant to freezing (Xylander and Neuermann 1990). Xylander (1991) has reviewed current knowledge of myriapod immune defence reactions. Responses to parasites are covered in Section 9.2. Osmoregulation of the blood is dealt with in Section 5.6.

Haemocyanin has been detected recently in the blood of the centipede *Scutigera coleoptera* (Mangum *et al.* 1985). However, the presence of haemocyanin in the blood of millipedes has yet to be confirmed.

5.4 Excretory processes

5.4.1 Introduction

Around 80–90 per cent of the food ingested (on a dry weight basis) by millipedes is voided as faeces. By far the most important site for assimilation of material is the midgut epithelium (Fig. 4.3). Products of digestion pass through the peritrophic membrane (Fig. 4.6) and are assimilated

---

Fig. 5.5 (A) The Kalahari millipede *Triaenostreptus triodus* (Spirostreptidae) showing sampling and injection site. (B) Nutrient agar seeded with 10^5 bacteria *E. coli* per ml. (i) Saline-injected haemolymph sample (28 μl haemolymph). (ii) Haemolymph sample (30 μl haemolymph) from a millipede 48 hours after injection of 10^7 bacteria *E. coli* per gram. Loaded 1.5 mg total protein in each well. (C) (a), *E. coli* gel overlay. Zones of antibacterial activity against gram-negative *E. coli* are evident in wells 1, 2, 4, and 5 (arrowed). Antibacterial activity was absent in the control haemolymph sample (well 3). (b), duplicate gel of (a), Page Blue and silver stained. Wells 1, 2, 4, and 5 are haemolymph samples from two millipedes 48 hours after injection of 10^7 bacteria *E. coli* K12 per gram. Well 3 is control haemolymph from a non-injected millipede. The direction of protein migration is from top to bottom. Reproduced from Van der Walt *et al.* (1990) by kind permission of the authors and Springer-Verlag.

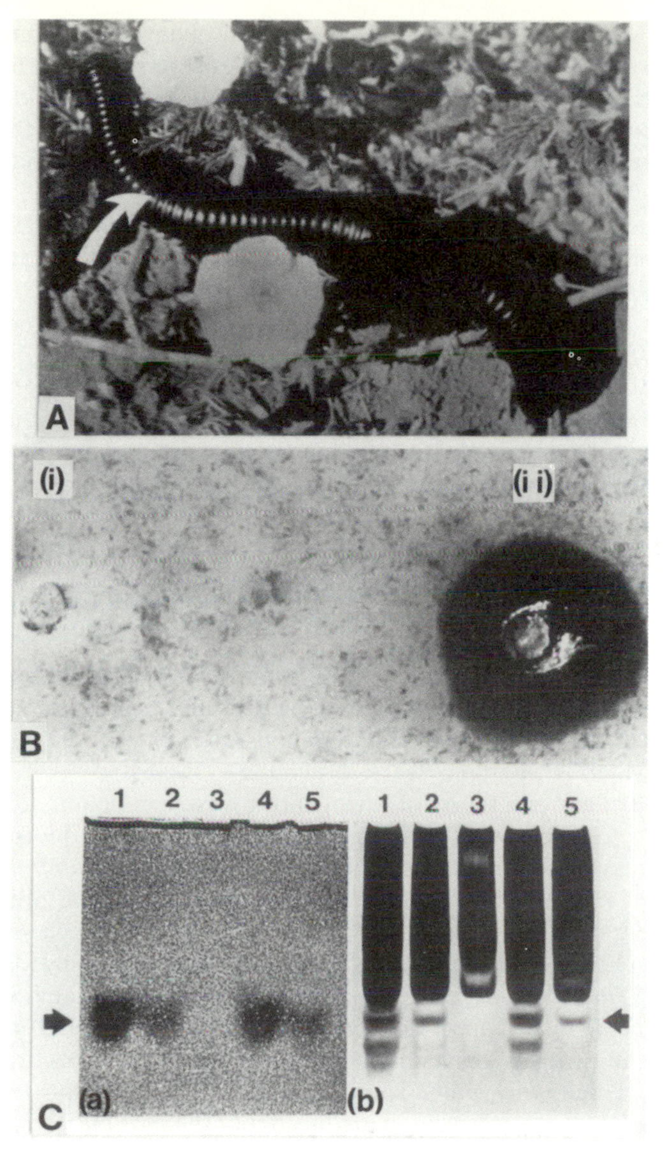

across the microvillus border of the cells. These products have to be small to get through the pores in the peritrophic membrane. There is no evidence that food material is phagocytosed directly by the cells of the midgut.

Once the material has entered the cells of the midgut, further digestion takes place. Products of this intracellular digestion pass across the basement membranes of the midgut cells into the liver cells (Fig. 4.3) and may be transported subsequently around the body to the organs that are bathed in blood. Unwanted products of digestion have to be removed from the circulation. These can be excreted, either by storage in an inert form within the tissues (liver cells, fat body), or removed entirely by loss to the external environment via an organ that either opens directly through the external cuticle (nephridial organs) or into the lumen of the gut (midgut, Malpighian tubules).

Numerous waste products are produced during feeding, digestion, and metabolism. Two of these will be covered in detail to demonstrate the main principles of excretion.

5.4.2 Nitrogenous waste

One of the main excretory products of animals is nitrogenous waste derived from the breakdown of proteins and their constituent amino acids. This waste nitrogen can be excreted in three main forms: ammonia, urea, and uric acid. Ammonia is the simplest form but is highly toxic and soluble. It cannot be stored in the body and must be excreted rapidly because it is a strong metabolic poison. Aquatic animals can produce ammonia as their principal nitrogenous excretory product because they are able to excrete it directly into the water.

Terrestrial animals must conserve water and convert nitrogenous waste into a relatively insoluble form if it is to be stored for any length of time before excretion. These forms are urea and uric acid. Uric acid can be crystallized and forms the main nitrogenous excretory product of birds and spiders (which is why their faeces are white).

In millipedes, two forms of nitrogenous waste predominate: ammonia (which has to be excreted rapidly) and uric acid (the stored form). Urea is detected less frequently and in a study on *Cylindroiulus londinensis*, Bennett (1971) could not detect urea at all in the faeces. The proportion of non-protein nitrogen as ammonia or uric acid varies between species. In a study by Hubert and Razet (1965) of total non-protein nitrogen in the faeces of *Glomeris marginata*, 40 per cent was ammonia and 33 per cent was uric acid. In *Cylindroiulus londinensis* however, only 20 per cent of non-protein nitrogen was ammonia whereas 70 per cent was uric acid. No urea was detected in either species. Anderson and Ineson (1983) were unable to detect any uric acid in the faeces of *Glomeris marginata* although the animals themselves contained 2.5 per cent uric acid by weight. They pointed out that figures for ammonium ion concentrations in the faeces may be considerably underestimated because much of it may be assimilated by bacteria in the lumen of the hindgut.

Uric acid is stored by most species as large spherical bodies of potassium urate in the fat body (Hubert 1979*a,b*). Anderson and Ineson (1983) have suggested that this might form a reserve of nitrogen for use when food contains insufficient amounts of the element.

5.4.3 Metals

Millipedes, like all animals, must assimilate sufficient quantities of essential 'trace' metals, such as copper, iron, and zinc, for a wide range of biochemical processes in the body. However, because the concentrations of these metals in the food are unpredictable, there are occasions when more is assimilated than is needed. In addition, non-essential elements such as cadmium and lead may pass into the midgut cells along routes normally taken by essential elements. In metal-polluted sites (see Section 10.7), concentrations of essential and non-essential metals in leaf litter may be more than an order of magnitude greater than in uncontaminated sites.

Invertebrates are able to detoxify these metals by storage in an insoluble form inside the cells in so-called 'granules' (Simkiss 1976; Taylor and Simkiss 1984). In the most recent review of this topic, Hopkin (1989) proposed that there were three main types of intracellular metal-containing granule.

1. Type 'A' granules are concentrically structured and are composed predominantly of calcium and magnesium phosphates. They may also contain zinc and lead.
2. Type 'B' granules are homogeneous in appearance and contain sulphur in association with metals such as copper and cadmium.
3. Type 'C' granules are also homogeneous in appearance and consist of waste iron, probably in the form of haemosiderin.

All types probably occur in millipedes although they can be difficult to recognize as such because once formed, the granules may coalesce in large excretory vacuoles thus confusing their original chemical composition.

Hubert (1977, 1978*a,b*, 1979*a,b*) was the first researcher to identify the sites of metal storage in millipedes. A wide range of metals were detected in granules in the midgut, liver, fat body, and Malpighian tubules of several species. More recently, Köhler and Alberti (1991) have identified numerous type A granules containing calcium phosphate in the cells of the midgut (Fig. 4.4). Millipedes that have been fed a lead-contaminated diet accumulate the metal in these granules. Preliminary observations on the dynamics of zinc, cadmium, and lead in *Glomeris marginata* were conducted by Hopkin *et al.* (1985) (see Fig. 10.7).

The granules may not function entirely as sites of storage-detoxification. Calcium-containing granules occur in the ovaries and eggs of several species and may supply calcium for the initial calcification of the cuticle of millipedes before they begin to feed (Crane and Cowden 1968; Hubert 1975; Petit 1970).

5.5 Organs involved in excretion

Seifert (1979) produced an excellent review on the evolution of excretory organs in terrestrial arthropods, with special reference to myriapods. He recognized ten organs which can be considered in the widest sense to be excretory in some or all of their functions.

5.5.1 Midgut epithelium

The midgut epithelium (Fig. 4.3) may act as a temporary store for some excretory products. These could be derived from the liver cells, or could be material which has been retained following assimilation from the lumen of the gut. Excretory bodies and metal-containing granules can be excreted by lysis of the cells into the lumen and hence lost via the faeces.

5.5.2 Liver

The cells of the liver may also accumulate excretory products. Some liver cells break down and release their contents into the blood for transport to other storage or excretory organs. Thus, they are analogous to the chloragogenous tissues of earthworms.

5.5.3 Integument

The integument and associated tissue (Fig. 3.9) may contain end-products of metabolism. However, care should be taken to determine whether the material is in the cuticle itself, or in the underlying tissues. It is an attractive idea that unwanted material could be stored in the cuticle in an inert form to be lost at moult. However, since most millipedes eat their exoskeleton after ecdysis, this would not be an effective method of permanent excretion.

The underlying tissues may contain a variety of metabolic end products which can serve a useful role as pigments (e.g. melanin, ommochromes, carotenids). In some species, concretions of uric acid are stored here, which appear white through the cuticle. The sub-cuticular tissues are a major site for the storage of zinc (Hopkin *et al.* 1985; Köhler and Alberti 1991).

5.5.4 Exocrine glands

Exocrine glands, such as the salivary glands, produce secretions which may contain unwanted material, although this is not likely to be a major excretory route. Material excreted in this way has to be insoluble in the digestive enzymes otherwise it will be reabsorbed during passage through the midgut. Glands which open onto the surface of the body, such as defence glands, may also contain waste material although again, this is probably not a major route of excretion.

5.5.5 Haemocytes

Haemocytes (covered in Section 5.3) may take up material from the blood (larger material by phagocytosis), and degrade and transport it around the body.

5.5.6 Nephridial organs

Nephridial organs are true excretory organs and are derived, as the name implies, from the characteristic osmoregulatory and excretory organs of annelids. Annelids have one pair in each segment but most arthropods (including millipedes) have only one pair (Hubert 1973). In millipedes, these open ventral to the mouthparts (El-Hifnawi 1973). Seifert (1979) provided a classic description of the evolution of these organs and the following account is based on his review.

The evolution of an exoskeleton by arthropods meant that the coelom, as a cavity for the retention of fluid for a hydrostatic skeleton, could be dispensed with. Thus in arthropods there is no closed circulation. All organs are bathed in blood which is kept moving between the various sinuses by the pumping action of the dorsal vessel or heart. Consequently, only a few nephridia are needed to clean up the blood, and in most arthropods (including millipedes) they have been reduced to one pair. These must be present from an early stage of development. Since millipedes are anemeric forms with post-embryonic teloblasty, they must be located anteriorly.

The passage of blood into the nephrostome of arthropods must be restricted in some way to stop essential metabolites from 'leaking out' into the excretory duct. To prevent this, the nephrostome is covered by a sacculus, which selects the substances that will reach the tubule (Fig. 5.6). The sacculus is formed from cells that surround a lumen into which primary urine is forced by pressure between pedicels (slits) in the basement lamina. Thus, the pedicels form a primary filter which retains cells and large molecules in the blood.

Much reabsorption takes place inside the sacculus before the secondary urine is allowed to pass down the excretory duct to the outside. The sacculus probably evolved from accumulations of nephrocytes (see below) at the open nephrostomes. There is little information on the composition of the urine that is released to the outside since the volumes involved are so small.

5.5.7 Nephrocytes

The nephrocytes are cells that take up substances from the haemolymph which are then metabolized. Some of the products of this metabolism are stored and some are returned to the circulation. In *Orthomorpha (Oxidus) gracilis*, Seifert and Rosenberg (1976) showed that nephrocytes were

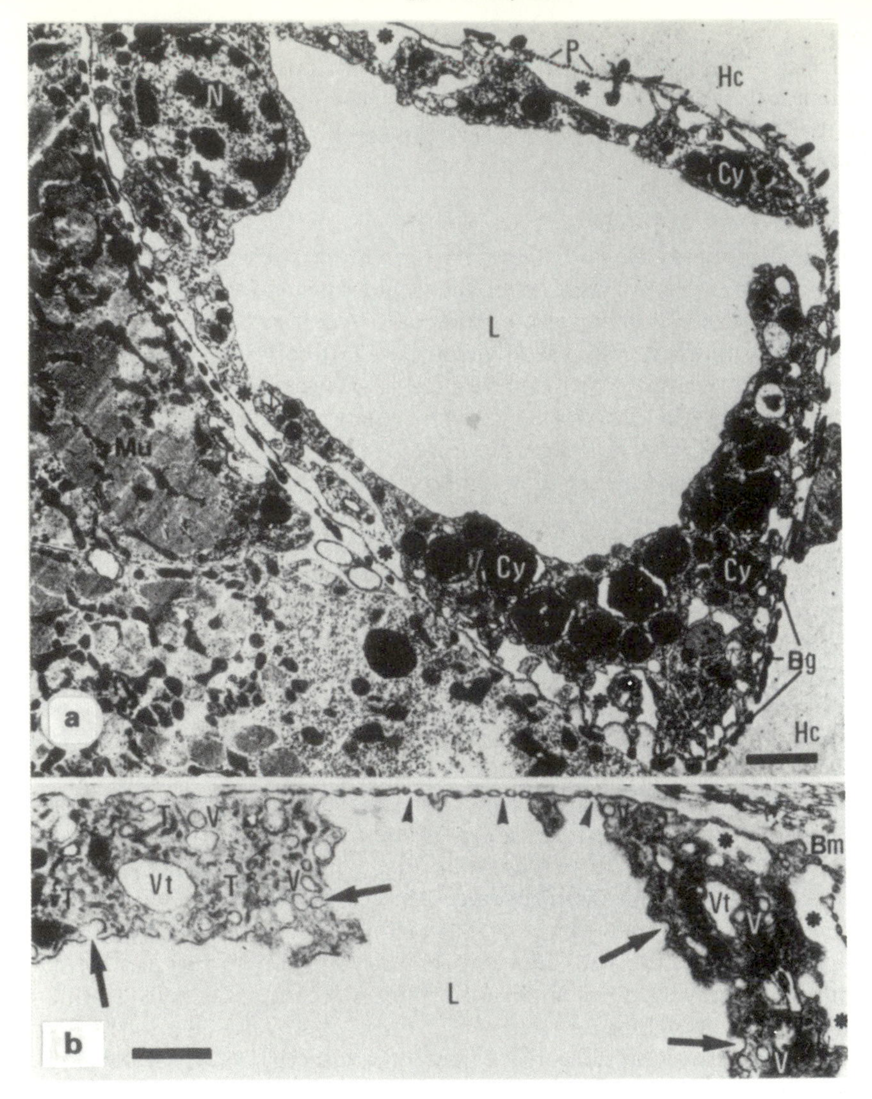

Fig. 5.6 Transmission electron micrographs of sections through the sacculus of the maxillary nephridium of *Polyxenus lagurus*. Bg, connective tissue; Bm, basement lamina; Cy, cytosome; Hc, haemocoel; L, lumen; Mu, apposed muscle; N, nucleus; P, pedicels; T, tubular structures; V, coated vesicles; Vt, transparent vacuole. The arrows in (b) point to vesiculations at the cell surface, and the arrow heads to diaphragms between neighbouring pedicels. The cortical labyrinth is indicated by asterisks. Scale bars = 2 μm (a) and 1 μm (b). Reproduced from Seifert (1979) by kind permission of the author and Academic Press.

present in the pericardial and perivisceral sinus of the trunk in all except the last two segments. The nephrocytes were present individually, or in bunches attached to the fat body, muscles, connective tissues, or tracheae. In this form, they present a large surface area for exchange of material with the blood.

5.5.8 Ecdysial glands

Seifert (1979) has suggested that ecdysial glands of arthropods should be treated as excretory organs. He gives the following justification. The metabolism of arthropods is incapable of forming the sterol ring system and thus synthesizing substances such as cholesterol. Therefore, all arthropods need sterols as essential nutrients. Since (like vertebrates) they can not degrade them, cholesterols accumulate within the body and must be removed. This is accomplished by cells which synthesize ecdysteroids from them, i.e. the ecdysial glands. Hence these are also, in the widest sense, excretory glands.

Only after mutations had enabled certain nephrocytes to synthesize moulting-stimulating hormones from cholesterol, could the formation of the arthropod cuticle be of selective advantage.

In centipedes, the ultrastructure of the ecdysial organs is almost identical with nephrocytes, and they are almost certainly derived from them. Ecdysial glands have yet to be positively identified in millipedes. In *Polyxenus lagurus*, they may be situated in the perioesophageal glands.

5.5.9 Malpighian tubules

All millipedes possess a pair of Malpighian tubules. These are blind-ending with their open ends connected to the gut at the midgut–hindgut junction. Seifert (1979) discussed their embryological origins and suggested that they may have evolved from nephrocytes.

Malpighian tubules evolved with the colonization of drier habitats by arthropods. In millipedes, they are primarily organs of nitrogenous excretion although several other molecules may also be lost via this route. Large numbers of granules containing uric acid and metals are accumulated within the cells and are discharged into the lumen for excretion in the faeces (Hubert 1972, 1979*a,b*). Mechanisms exist for water reabsorption in regions of the tubule, and in the hindgut.

The structure and function of Malpighian tubules have been studied most extensively in *Glomeris marginata*. These are probably typical of most millipedes and are described below. The special adaptations of the hindgut–Malpighian tubule complex of *Polyxenus lagurus*, which lives in dry habitats, are examined at the end of this section.

Considerable differences exist between the fluid-secreting upper part and the lower part of the Malpighian tubules of *Glomeris marginata*. The upper tubule has a high permeability to compounds of high molecular weight (Johnson and Riegel 1977*a*). The epithelium contains a very extensive system of intercellular spaces which are linked directly to junctions that are specialized, apparently, to provide a low-resistance extracellular pathway between the blood and the tubule lumen. Ferritin, iron dextran, and horseradish peroxidase can cross the basal lamina and enter the tubule lumen of the upper segment and are accumulated within intracellular vesicles (Johnson and Riegel 1977*b*).

Farquharson (1974*a,b,c*) made an extensive study of the physiology of isolated Malpighian tubules of *Glomeris marginata*. A Ringer solution was developed which would support secretion of fluid *in vitro*. Isolated tubules produced fluid that was identical to the bathing medium with regard to ion concentrations and osmotic pressure. Tubules secreted normally in a potassium-free Ringer, but ceased to function in a sodium-free Ringer, indicating that the sodium ion is the 'prime mover'. The hypothesis that a sodium pump is responsible for fluid secretion was confirmed by the findings that the rate of fluid production depended on the availability of sodium ions, and that secretion was inhibited by the metabolic inhibitors 2,4-DNP, azide, and cyanide. Parts of the tubule appear to act as a molecular filter, i.e. molecules move through the tubule wall in inverse relation to their size (Farquharson 1974*c*).

The Malpighian tubules of *Polyxenus lagurus* are associated closely with the hindgut (Fig. 5.7; Schlüter and Seifert 1985*a*). The tubules are ensheathed by an envelope that consists of several flattened cells. This complex resembles the cryptonephric system in insects. Four segments of the tubule can be recognized:

(1) the distal segment, joined by a short transition zone to
(2) the thin segment within the perinephric compartment. This leads into
(3) the thick meandering segment which has no microvilli. This is joined to
(4) the proximal segment, which discharges into the hindgut.

The complex creates osmotic gradients that promote resorption of water from the rectum (Fig. 5.8).

5.5.10 Fat body

The fat body is a diffuse organ that lies in the body cavity of millipedes. It is bathed in blood and functions primarily as a storage organ for lipids, glycogen, proteins, and uric acid (Subramoniam 1972). The adipocytes incorporate a variety of granules that contain metals or potassium urate (Hubert 1974, 1975, 1978*a,b*, 1979*a,b*) and are sites of permanent storage-excretion of unwanted substances.

Care should be taken when the importance of this organ in metal storage is being examined. It is essential to separate the fat body from the tissues underlying the exoskeleton. For example, Hopkin *et al.* (1985) showed that

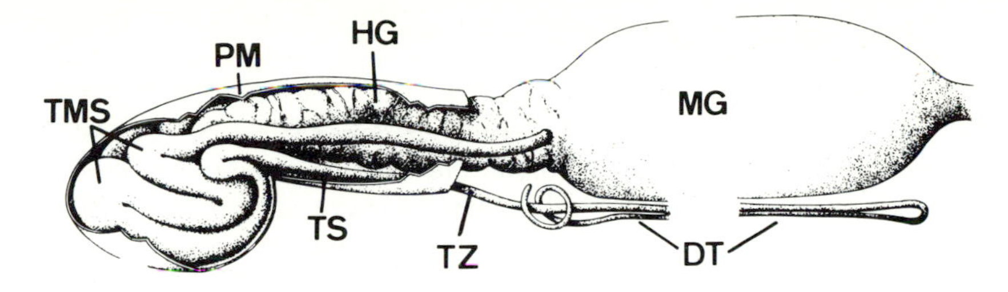

Fig. 5.7 Hindgut–Malpighian tubule complex in *Polyxenus lagurus*. Semi-schematic diagram of the gut and Malpighian tubules. Gut: HG, hindgut; MG, midgut. Malpighian tubules: DT, distal tubule segment; PM, perinephric membrane; TMS, thick meandering segment; TS, thin segment; TZ, transition zone. Reproduced from Schlüter and Seifert (1985*a*) by kind permission of the authors and Artis Bibliotheek, Amsterdam.

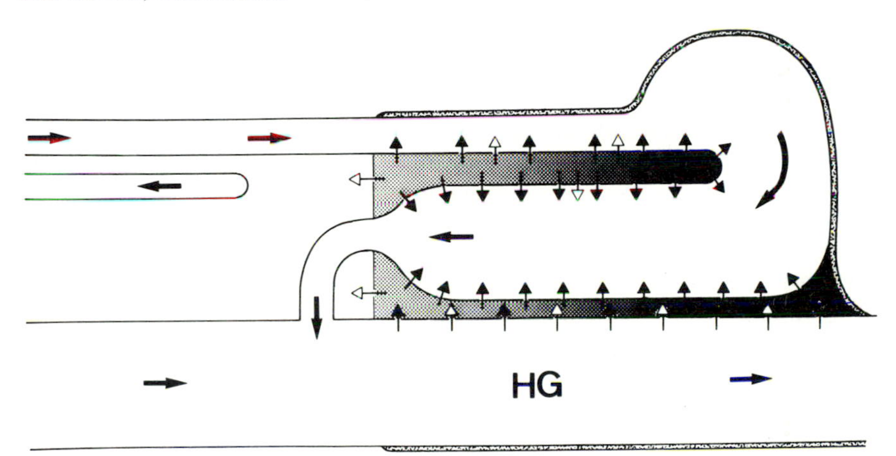

Fig. 5.8 Hindgut–Malpighian tubule complex in *Polyxenus lagurus*. Scheme of the assumed mechanism of ion- and water-recycling in the hindgut region. The solid arrows indicate ion movements, the empty arrows indicate water movements. HG, hindgut. Reproduced from Schlüter and Seifert (1985*a*) by kind permission of the authors and Artis Bibliotheek, Amsterdam.

the fat body of *Glomeris marginata* was relatively unimportant in the storage of zinc, cadmium, and lead in comparison to the midgut and subcuticular tissues.

Subramoniam (1971) showed that the fat body was capable of accumulating proteins from the blood. He injected horseradish peroxidase into the haemocoel of *Spirostreptus asthenes* and demonstrated that the marker could be detected subsequently in the fat body. The uptake of markers by the fat body of millipedes has been studied also by Oudejans (1972*a*) and Oudejans and Zandee (1973).

5.6 Osmoregulation: water and ionic balance

5.6.1 Introduction

The percentage water content of a millipede depends on the species and the physiological state of the animal in question. Typically, it is about 60 per cent (Appel 1988). Most millipedes are more susceptible to losing water than insects (Mantel 1979). Nevertheless, some species can tolerate loss of about a third of their body water, providing this is temporary (Meyer and Eisenbeis 1985).

In general, big millipedes are less susceptible to water loss than small ones (Crawford 1979). There are also considerable differences between species that are related to environmental factors (see Table 5.5). Crawford *et al.* (1986, 1987) provide more detailed coverage of this topic than is possible here.

Millipedes have evolved a variety of behavioural and physiological mechanisms to reduce water loss. These have been summarized by Crawford (1979) and are described below.

For a soil-bound millipede, dryness beyond a certain point sets up a gradient between the animal and the air which is osmotically unfavourable. This will result in progressive dehydration and eventual death unless an appropriate combination of the following compensatory mechanisms occurs.

(1) walking to a wet area;
(2) minimization of cuticular, respiratory, and other forms of water loss (including rolling up);
(3) elevation of internal osmotic pressure; or
(4) addition of water by means of metabolism or transport from the external environment.

Some species of millipedes have developed water conservation to such a degree that they have been able to colonize very dry habitats including deserts. Their cuticular permeability is much lower than species which live in moist habitats. For example, the cuticular permeability of the desert living *Orthoporus ornatus* is about 8 $\mu g\,cm^{-2}\,h^{-1}\,mmHg^{-1}$ whereas for *Oxidus gracilis* it is about 80 $\mu g\,cm^{-2}\,h^{-1}\,mmHg^{-1}$ and for *Glomeris marginata*, about 200 $\mu g\,cm^{-2}\,h^{-1}\,mmHg^{-1}$ (Appel 1988—see also Edney 1951 for early work on *Glomeris*). *Polyxenus lagurus*, which has been extremely successful in colonizing dry habitats, is able to restrict transpiration rates at quite low relative humidities (Eisenbeis and Wichard 1987).

5.6.2 Osmoregulation by behavioural means

The most obvious ways in which millipedes can reduce water loss are to either avoid dry areas, or reduce the surface area that is exposed to the atmosphere.

Table 5.5 Rates of water loss in millipedes from various habitats at 0% relative humidity at two different temperatures. Reproduced from Meyer and Eisenbeis (1985) by kind permission of the authors and Artis Bibliotheek, Amsterdam

	Habitat (alt.)		15°C		25°C	
			Permeability (± SE)	Resistance*	Permeability (± SE)	Resistance
		n	$\mu g\ cm^{-2}\ h^{-1}\ mmHg^{-1}$	$s\ cm^{-1}$	$\mu g\ cm^{-2}\ h^{-1}\ mm\ Hg^{-1}$	$s\ cm^{-1}$
Enantiulus nanus	oak litter (670 m)	3	53.4 (± 4.0)	63.7	98.1 (± 25.1)	34.7
Mastigona mutabilis	oak litter (670 m)	3	234.8 (± 10.0)	14.5	284.8 (± 55.9)	11.9
Leptoiulus saltuvagus	alder litter (2000 m)	7	53.7 (± 13.8)	63.3	55.5 (± 6.8)	61.3
Haasea fonticulorum	alder litter (2000 m)	3	185.0 (± 11.7)	18.4	236.9 (± 13.9)	14.4
Ochogona caroli	alder litter (2000 m)	6	169.8 (± 17.7)	20.0	147.0 (± 17.2)	23.1
Trimerophorella nivicomes	high alpine grassland (2500 m)	3	18.3 (± 1.8)	185.9	45.9 (± 9.5)	74.1

* Resistance ($s\ cm^{-1}$) is the reciprocal of permeability ($cm\ s^{-1}$). A quick rule of thumb for the conversion of $\mu g\ cm^{-2}\ h^{-1}\ mmHg^{-1}$ to $cm\ s^{-1}$ is given by the relationship: $2.94 \times (\mu g\ cm^{-2}\ h^{-1}\ mmHg^{-1}) = (cm\ s^{-1}) \times 10^4$ (Edney 1977).

Most millipedes exhibit a strong preference for areas of high humidity. Toye (1966*a*) showed that the Nigerian millipedes *Oxydesmus* (=*Coromus*) sp. and *Habrodesmus falx*, always migrated to areas with high humidity. This response was in preference to their normal photopositive behaviour. In moist conditions, they always moved towards light. *Ommatoiulus moreleti* in Australia are able to survive summer dry spells by crawling deep into grass tussocks where humidity is higher (Baker 1980). Peitsalmi (1974) showed that *Proteroiulus fuscus* always migrated to the region of highest humidity irrespective of whether this was in the upper or lower end of a vertical humidity gradient.

Many species are most active at night, when humidity is usually highest (Toye 1966*b*). Millipedes in areas subject to seasonal drought may survive the dry period by becoming dormant (Crawford 1979). Some remain in moulting chambers protected from desiccation for lengthy periods (Lewis 1971*a,b*, 1974). Indeed, moulting is a time when millipedes are vulnerable to desiccation *and* waterlogging (Miller 1974), so the construction of a moulting chamber insulates them from wide variation in humidity (Fig. 8.6).

In order to reduce the surface area of the body exposed to the atmosphere, many millipedes roll up when conditions are dry. Cylindrical millipedes coil up to form a spiral (Fig. 3.12(b)) whereas pill millipedes such as *Glomeris marginata* are able to roll into a sphere so that only the dorsal surface of the body is exposed (Fig. 3.14(a)). Where millipedes are present in large numbers, aggregating behaviour may reduce the collective surface area exposed to the air (Toye 1966*c*).

5.6.3 Osmoregulation by physiological means

There are two ways in which a millipede can respond to the increased osmolality of the blood due to dehydration. One way is to tolerate the lower osmolality, the other is to regulate haemolymph osmolality (Fig. 5.9). These are the strategies adopted by desert scorpions and beetles respectively (Riddle *et al.* 1976). The desert millipede *Orthoporus ornatus* displays both osmotic regulation and tolerance depending on sex and on duration of desiccation.

Millipedes from temperate climates are not tolerant to rapid changes in the osmolality of the haemolymph. Thus, the ionic composition of the blood must be regulated within fairly narrow boundaries. Most British diplopodologists are familiar with the collections of dried-up julids that occur under loose bark during the summer. Presumably these millipedes were unable to migrate to a damp place and died through dehydration. In *Pachydesmus crassicutis*, haemolymph osmolality increases by a greater amount when the animal is dehydrated rapidly (165–208 mosmol) than when it is dehydrated more slowly (165–182 mosmol) (Woodring 1974).

Sodium and potassium ions are the main contributors to total osmolality of the blood of millipedes. In *Julus scandinavius*, the mean concen-

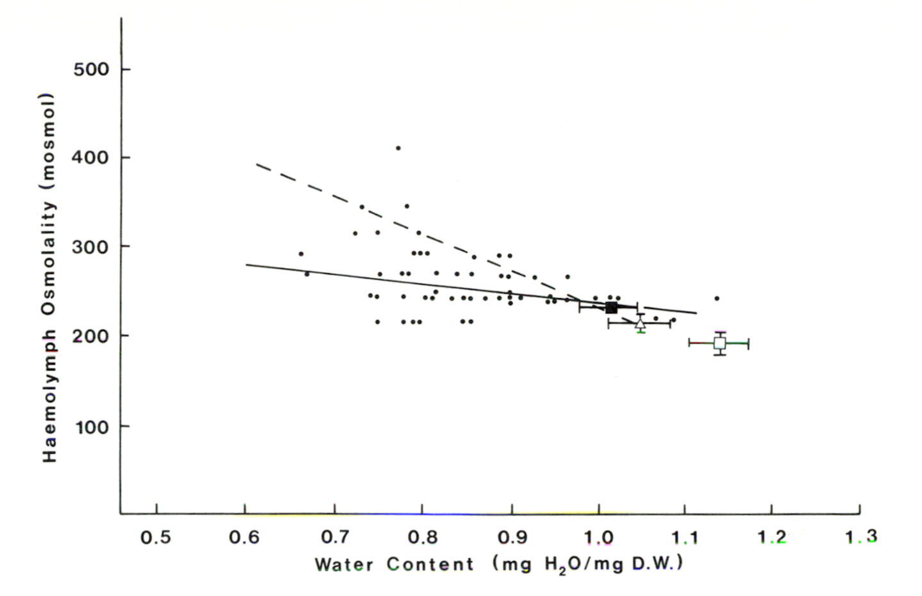

Fig. 5.9 Influence of dehydration and rehydration on the water content and haemolymph osmolality of male *Cylindroiulus londinensis*. The solid line is a regression fitted to the points. Water content and haemolymph osmolality values are presented for field collected (■), pretreatment (△), and rehydration (□) groups. The dashed line depicts a predicted osmotic response of pretreatment animals to dehydration in the absence of osmoregulation. Vertical and horizontal bars are 95 per cent confidence limits. Limits of the mean osmolality value for field animals are less than the symbol size. Redrawn from Riddle (1985) by kind permission of the author and Pergamon Press.

tration of sodium is 70 $\mathrm{mM\,l^{-1}}$ and of chloride, 61 $\mathrm{mM\,l^{-1}}$ (Sutcliffe 1963).

The permeability of the cuticle of millipedes is related to some extent to the dryness of the habitat they inhabit (see above). In the desert species *Orthoporus ornatus*, the epicuticle is waterproofed by the presence of a surface wax layer (Crawford 1979). Spirobolid and polydesmid millipedes do not appear to possess a spiracular closing mechanism similar to insects, although the openings may be occluded by a combination of diplosegmental overlap and close appression of the coxae (Stewart and Woodring 1973).

Eisenbeis and Wichard (1987) found that *Polyxenus lagurus* was able to increase its water content rapidly, but only in the early morning between 5.00 and 6.00 a.m. Ambient moisture levels are high at this time because of dew formation. Crawford (1978, 1979) has suggested that some millipedes may be able to take up water actively from the air across the cuticle. A more likely route is via hyperosmotic fluid on a moist surface. This is the strategy used by terrestrial isopods which can absorb water directly from the atmosphere via hyperosmotic fluid on the ventral pleonites (Wright and Machin 1990). The site of water vapour uptake in millipedes is unknown but is likely to be the rectum (see below).

Species that live in dry conditions have also evolved ways to reduce water loss via the faeces. *Polyxenus lagurus* has evolved a hindgut–Malpighian tubule complex for water reabsorption from the faeces (Section 5.5.9). All millipedes are probably capable of absorbing water across the epithelium of the hindgut (Crawford and Matlack 1979). Indeed, water and salt balance probably involves fairly uninhibited loss of fluids and ions from the Malpighian tubules, then selective reabsorption by the hindgut.

Species that live in dry conditions have drier faeces than species from moist environments. For example, the faeces of the hygric species *Pachydesmus crassicutis* contain 65–85 per cent water whereas those of the mesic *Orthoporus texicolens* contain 28–45 per cent water (Stewart and Woodring 1973).

The hindgut has the characteristic appearance of a transporting epithelium, with numerous intercellular channels and mitochondria (Schlüter 1980*d*). The isolated hindgut of *Orthoporus ornatus* is capable of active transport of sodium from the lumen side to the blood side, and of potassium from the blood to the lumen (Moffett 1975).

In addition to direct drinking, some species are able to rapidly redress water deficits by everting their rectum and pressing it against a moist substrate (Fig. 5.10; Crawford 1972). In an experiment on five alpine species of millipedes, Meyer and Eisenbeis (1985) showed that dehydrated *Mastigona mutabilis* could absorb 79 per cent of their original water content via an everted rectum in one hour (Table 5.6).

Table 5.6 Rates of water loss and highest observed rates of water uptake from moist filter paper in six species of millipedes at 0% relative humidity and 25 °C. Both loss and uptake rates are expressed as percent of original water (m_0). Reproduced from Meyer and Eisenbeis (1985) by kind permission of the authors and Artis Bibliotheek, Amsterdam.

	Water loss (± SE)		Water uptake	
	$-\Delta m_0$ (% h^{-1})	*n*	$+\Delta m_0$ (% h^{-1})	*n*
Leptoiulus saltuvagus	9.1 (± 0.3)	4	50.1	5
Enantiulus nanus	17.4 (± 5.8)	3	19.0	5
Trimerophorella nivicomes	9.8 (± 2.1)	3	64.2	4
Mastigona mutabilis	54.6 (± 10.2)	3	79.4	5
Haasea fonticulorum	65.5 (± 3.6)	2	19.2	4
Ochogona caroli	44.6 (± 6.5)	4	57.6	5

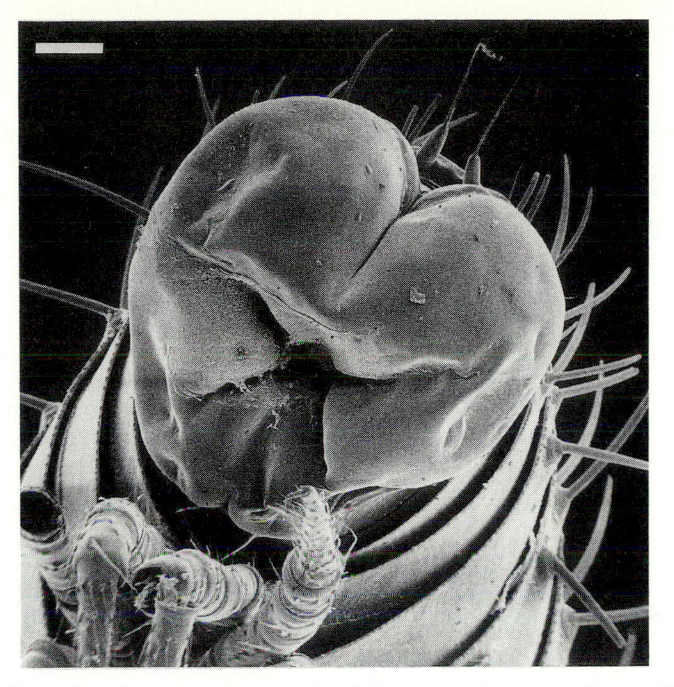

Fig. 5.10 Scanning electron micrograph of the everted rectum of a stadium VIII female of *Mastigona mutabilis*. Scale bar = 100 μm. Reproduced from Meyer and Eisenbeis (1985) by kind permission of the authors and Artis Bibliotheek, Amsterdam.

Females of the desert species *Orthoporus ornatus* may even resorb oocytes during periods of extreme dehydration stress (Crawford & Warburg 1982).

5.7 Thermoregulation

Physical methods of thermoregulation are not available to millipedes (i.e. diverting blood flow to the sub-cuticle to lose heat). Regulation of body temperature is achieved by migrating to areas of preferred temperature. Some desert millipedes deliberately bask in full sunlight, possibly to increase the efficiency of digestion (Crawford *et al.* 1987). In the deserts of New Mexico and Texas, *Orthoporus ornatus* remains at the surface until temperatures reach 35°C. It then seeks refuge under stones and in the centre of shrubs until the temperature drops again (Wooten *et al.* 1975).

6

Nervous, sensory, and neurosecretory systems

6.1 Nervous system

The sensory receptors and organs, and neurosecretory system of millipedes are among the better-studied areas of myriapod biology. These topics are covered in detail in this chapter in Sections 6.2 and 6.3. However, apart from anatomical descriptions (e.g. Newport 1843), little is known of the fine structure and function of the nerves of millipedes. Direct recording of nerve impulses has rarely been attempted. In the few cases where the ultrastructure of the nerves has been examined, they have been shown to be structurally similar to those of other terrestrial arthropods (Saita and Candia-Carnevali 1978). More research is required on this topic. Some of the more contentious questions concerning myriapod (and arthropod) phylogeny may be solved at the ultrastructural, rather than the anatomical, level.

6.2 Sensory systems

6.2.1 Introduction

Despite being one of the better-studied areas of millipede biology, we have still some way to go until the structure and function of the sensory receptors and organs are as well-understood as those of insects (Altner and Prillinger 1980; McIver 1975; Zacharuck 1985). Haupt (1979) has provided the most comprehensive review of myriapod sense organs. Much of what follows in this section is based on his excellent study. Eisenbeis and Wichard (1987) provide numerous scanning electron micrographs of the external appearance of millipede sense organs, some of which are included in this chapter (Figs 6.3, 6.5, 6.6, 6.7a).

The earliest comprehensive study was conducted by Cloudsley-Thompson (1951) on *Paradesmus (Oxidus) gracilis* and *Blaniulus guttulatus*. He showed, amongst other things, that they had a preference for a temperature of 15°C which was not exhibited if the antennae were removed. Care should of course be taken in interpreting such experiments as animals which have undergone such drastic surgery are unlikely to act in a normal fashion.

Before moving on to examine the structure and function of sensory organs, it is worth noting that millipedes have a strong sense of direction. This is in addition to an ability to follow chemical trails such as those made by ants (Akre and Rettenmeyer 1968; Rettenmeyer 1962). Mittelstaedt *et al.* (1979) conducted experiments on a species of *Spirostreptus*. They showed that if the walking plane is tilted, millipedes in the dark adopt a compromise course between the ('idiothetic') tendency to continue going straight ahead, and the ('allothetic') tendency to walk uphill. When the arena was set back into the horizontal after a 25–30 cm walk on the inclined plane, the millipedes turned in a direction which was approximately equal to the initial one before the tilt.

Barnwell (1965) also conducted experiments on the angle sense of millipedes. He showed that *Trigoniulus lumbricinus*, when forced through a corridor containing an abrupt turn, tended to turn upon emergence from the corridor at an angle which was opposite and approximately equal to the angle of the forced turn. This response held for forced turns up to 120°. The millipedes were able to distinguish differences in the angle of forced turn of as little as 10°.

Haupt (1979) categorized millipede sense organs as either 'simple' or 'complex'. Simple organs are mechanoreceptors, gustatory receptors, and olfactory receptors. Complex organs are eyes, Tömösváry organs, and trichobothria. These will be examined in turn.

6.2.2 Simple sensory structures

6.2.2.1 Mechanoreceptors

Mechanoreceptors are usually innervated by one sensory cell and have a tubular body at the dendritic terminal. There is an articulation membrane and often a 'hair' which may be held in position by suspension fibres. In *Polyxenus lagurus*, the tubular body lies between the hair shaft and the socket septum (Fig. 6.1). The hair shaft provides leverage to displace the tubular body and the socket septum limits the extent of the tubular body's lateral displacement. At least one of these elements forms the basis for a structural bilateral symmetry along whose plane lies the direction of maximum receptor sensitivity (Gaffal *et al.* 1975).

6.2.2.2 Gustatory receptors

These receptors have an articulation membrane and often a mechanoreceptive dendrite at the base. Chemoreceptive dendrites extend into the lumen of the hair to a terminal pore. They respond to direct contact with chemicals, in contrast to olfactory receptors (discussed below) which respond to molecules in the air. Thus, sensilla of this type may have a dual function, mechanoreceptive and chemoreceptive (Nguyen Duy-Jacquemin 1985*a,b*).

Sakwa (1978) studied the two kinds of gustatory receptors on the antennae of *Tachypodoiulus niger*. The first type are large basiconic sensilla at

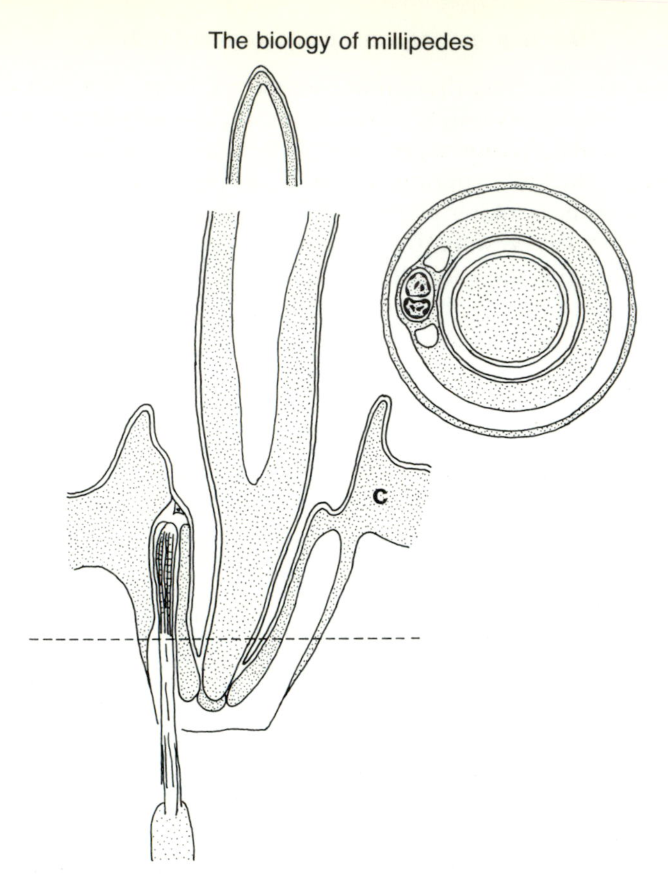

Fig. 6.1 Schematic diagram of a longitudinal and a transverse section through a mechanoreceptor of *Polyxenus lagurus*. The dotted line indicates the plane of transverse section. c, cuticle. Redrawn from Haupt (1979) by kind permission of the author and Academic Press.

the apex of the terminal article. The second type are smaller and are situated at the distal ends of articles 5,6, and 7. The distal end of the gnathocilarium also bears basiconic sensilla. These sensilla possess up to five receptor cells which respond electrophysiologically to water, 0.1 M sucrose + 0.01 M NaCl, 0.1 M glucose + 0.01 M NaCl, and 0.1 M NaCl alone. The sensilla of the gnathocilarium have a greater sensitivity than those on the antennae and function in determining the 'palatability' of potential food items (Sakwa 1974).

6.2.2.3 Olfactory receptors

These receptors come in a variety of shapes in arthropods. The functional cuticular surface is always perforated by numerous tiny pores. Below the cuticle are branching dendrites of numerous sense cells. They are of two main types: those with cuticular pore tubules, and those with spoke canals. This type of receptor probably also includes those which are sensitive to temperature and humidity. In millipedes, they are found in the Tömösvary organ (Section 6.2.3.2), and on the antennae (Section 6.2.4).

6.2.3 Complex sense organs

6.2.3.1 Eyes

The ability to detect light is of varying importance in the Diplopoda. Cavernicolous species, and those which live permanently in the soil (such as *Blaniulus guttulatus*) are blind. Those with eyes may respond positively or negatively to light. In Australia, the introduced species *Ommatoiulus moreleti* is attracted to the light that emanates from houses. This millipede causes a considerable nuisance when it enters dwellings in large numbers (McKillup and Bailey 1990). It has been suggested that the species may use the position of the moon in the sky for navigation and be confused by artificial lights (McKillup 1988).

Millipedes that possess eyes may have as few as one ocellus on each side of the head (e.g. *Stemmiulus*) or as many as 90 (e.g. *Dendrostreptus macracanthus*) (Enghoff 1990). However, the extinct *Glomeropsis*, which lived during the Upper Carboniferous, had up to 1000 ocelli in each eye (Kraus 1974).

Spies (1981) studied the structure of the eyes of *Polyxenus lagurus, Craspedosoma simile, Polyzonium germanicum*, and *Schizophyllum (Ommatoiulus) sabulosus* (Fig. 6.2). He concluded that whereas each ocellus of *Polyxenus* (Penicillata) could be derived from a single ommatidium (Fig. 6.2a), the ocellus of the three species of Chilognatha (Fig. 6.2b–d) could only be interpreted as a product of a merging of several associated ommatidia as a result of multiplication and re-arrangement of former ommatidial elements.

In *Glomeris marginata*, each ocellus consists of a corneal lens, and corneagenous and retinal cells (Bedini 1970). The retinal cell axons contain many neurotubules and unite with those of other cells to form the optic nerve. Munoz-Cuevas (1984) has reviewed photoreceptive structures in arachnids and myriapods.

6.2.3.2 Tömösváry organs

Tömösváry organs, (or post-antennal temporal organs as they are sometimes known), are paired structures found in most myriapods (Tömösváry 1883). In millipedes, the openings are visible as a circular, oval, or U-shaped pore on each side of the head between the base of the antennae and the eye (Fig. 6.3). Numerous suggestions have been made as to their function, including sound reception (Meske 1962). However, ultrastructural studies have indicated that they are probably olfactory.

Bedini and Mirolli (1967) made a detailed study of the ultrastructure of

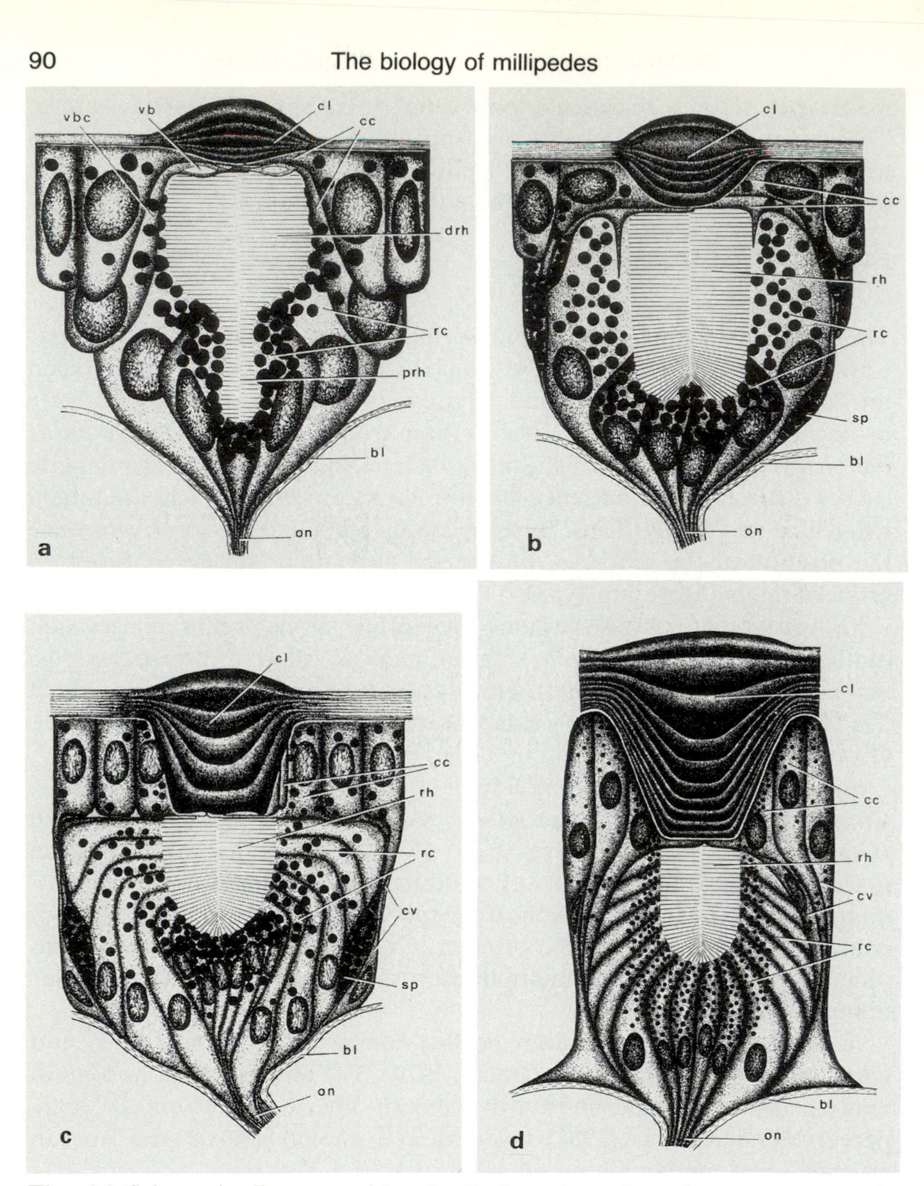

Fig. 6.2 Schematic diagrams of longitudinal sections through an ocellus of (a) *Polyxenus lagurus*, (b) *Polyzonium germanicum*, (c) *Craspedosoma simile* and (d) *Schizophyllum* (*Ommatoiulus*) *sabulosus*. bl, basal lamina; cc, corneagenous cells; cl, corneal lens; cv, covering cells; drh, distal rhabdom; on, optic nerve; prh, proximal rhabdom; rc, retinular cells; rh, rhabdom; sp, hypodermal cell processes with screening pigments; vb, vitreous body; vbc, vitreous body cell. Reproduced from Spies (1981) by kind permission of the author and Springer-Verlag.

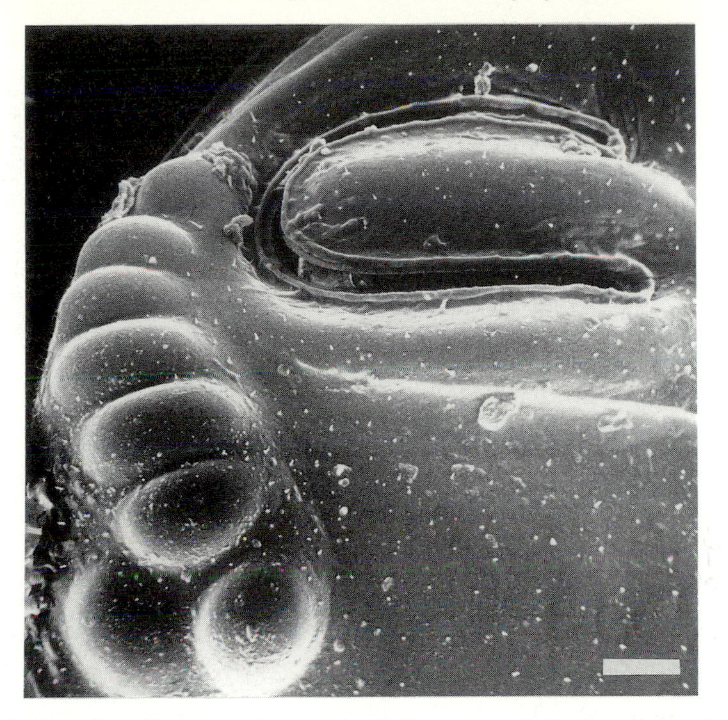

Fig. 6.3 Scanning electron micrograph of the eye and U-shaped opening to a Tömösvary (postantennal) organ of *Glomeris marginata*. Scale bar = 50 μm. Reproduced from Eisenbeis and Wichard (1987) by kind permission of the authors and Springer-Verlag.

the Tömösváry organs in *Glomeris marginata* (Fig. 6.3). The sensory epithelium is formed from a layer of primary sensory neurons and supporting cells which are covered by a porous cuticular plate. The dendrites of the neurons reach the surface of the epithelium. Their apical portion bears two cilia which pass through the pores of the cuticle and expand over this as a complex tubular structure. The tubules of the cilia are separated from the cavity of the organs (which is open to the outside) only by a thin lamella of amorphous material that is permeable to water. This supports the hypothesis that these organs are receptive to smell.

6.2.3.3 Trichobothria

Trichobothria are sensory hairs which are usually multi-innervated, have a directional sensitivity, and function primarily to detect slow air currents. In millipedes, they have been found only in the Penicillata (Fig. 6.4). Trichobothria have been studied extensively in *Polyxenus lagurus* by Tichy (1975) on whose account the following description is based.

Three sensory hairs are inserted at right angles to each other on opposite

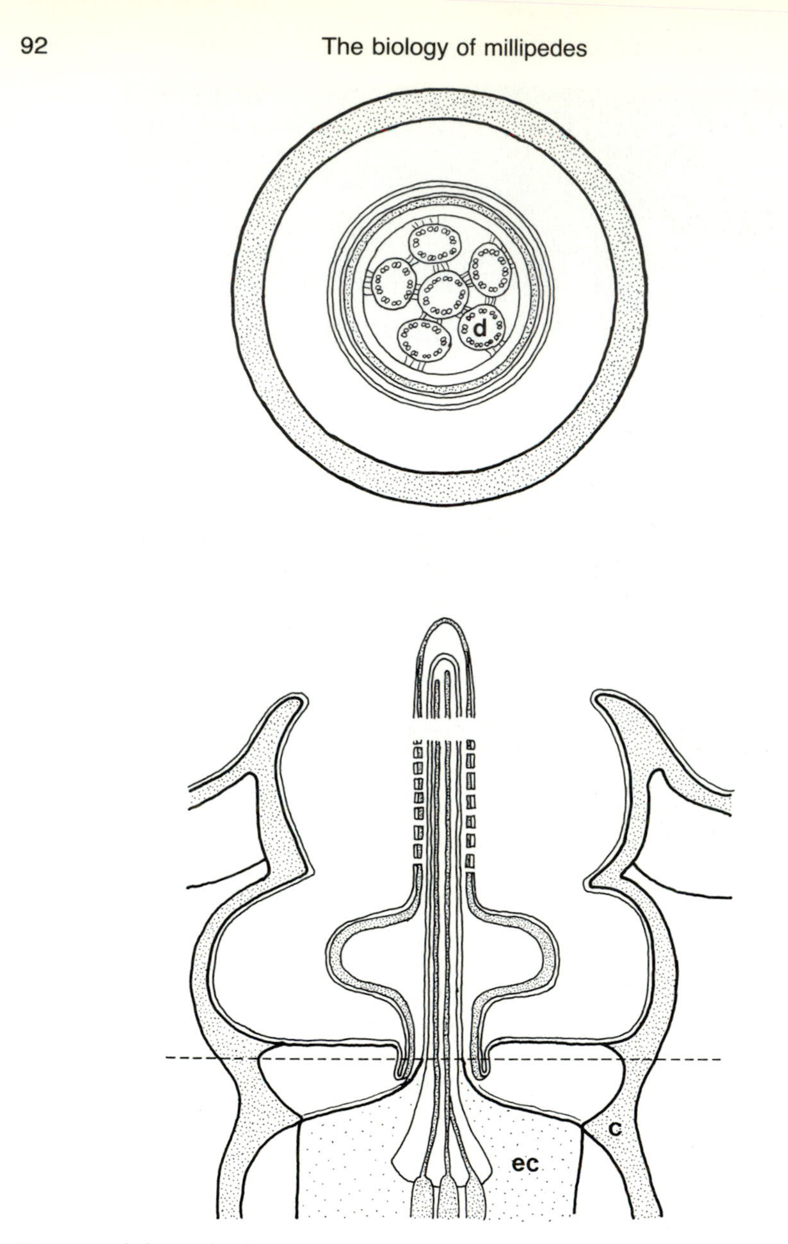

Fig. 6.4 Schematic diagram of a longitudinal and a transverse section through a trichobothrium of *Polyxenus lagurus*. The dotted line indicates the plane of transverse section. c, cuticle; d, dendrites; ec, envelope cells. Redrawn from Haupt (1979) by kind permission of the author and Academic Press.

sides of the head (Figs 6.5, 6.6). They display structural features characteristic of trichobothria, but also of insect-like scolopidia. These trichobothria possess five unusual features.

1. The dendritic cilia are enclosed within a capsule formed by enveloping cells.
2. The dendritic cilia are inter-connected by desmosome-like junctions.
3. The $9 \times 2 + 0$ organization of the dendritic ciliary microtubules is maintained over the entire length of the cilia.
4. There is no tubular body and the dendritic cilia are not swollen.
5. Pores and pore tubules occur in the lower halves of the hairs.

This uncommon combination of structural details makes it difficult to interpret the stimuli to which these organs respond. In addition to mechanoreception, the trichobothria of *Polyxenus lagurus* could possess several other functions including humidity and temperature determination.

Fig. 6.5 Scanning electron micrograph of the head region of *Polyxenus lagurus* (see also Fig. 6.6). Scale bar = 100 μm. Reproduced from Eisenbeis and Wichard (1987) by kind permission of the authors and Springer-Verlag.

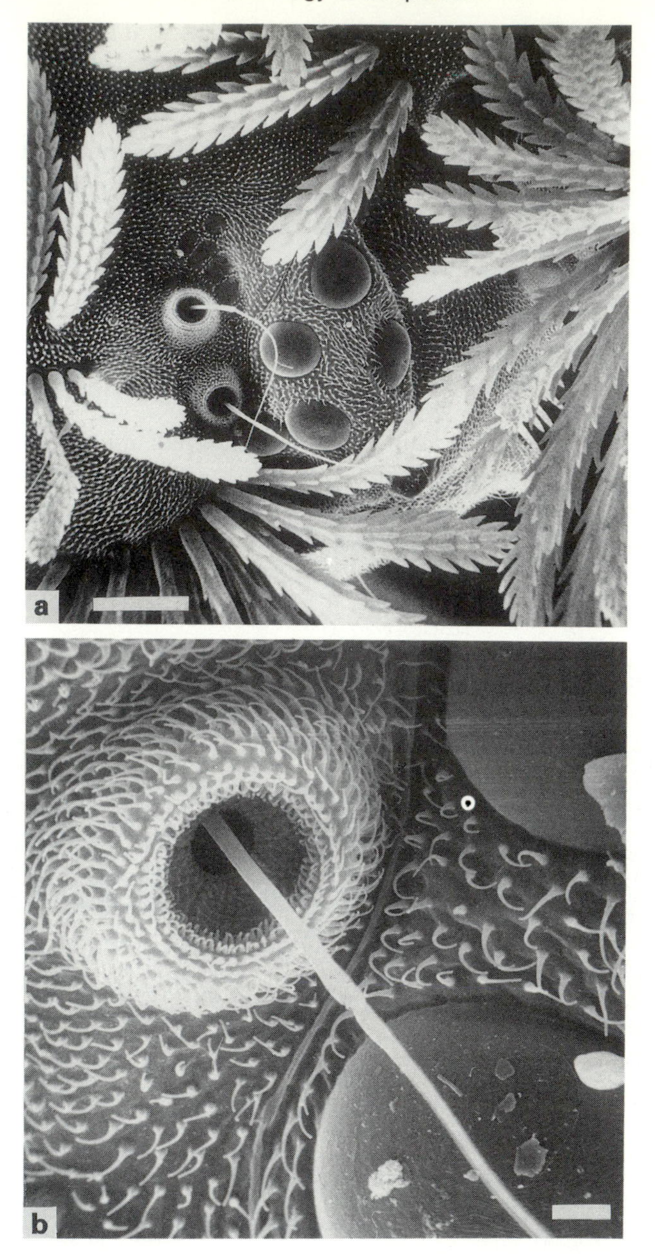

Fig. 6.6 Scanning electron micrographs of *Polyxenus lagurus*. (a) Close up view of the animal's left eye (see Fig. 6.5). Two of the three trichobothria can be seen to the left of the ocelli. The third is obscured behind a serrated setum to the left of the visible trichobothria. Scale bar = 25 μm. (b) Cupulate socket of a trichobothrium. Scale bar = 3 μm. Reproduced from Eisenbeis and Wichard (1987) by kind permission of the authors and Springer-Verlag.

6.2.4 Antennae

Anyone who has watched a millipede tapping its antennae on the substrate in front of it as it walks along will appreciate the importance of these structures in providing millipedes with information on what they are about to bump into! The importance of the antennae to millipedes was summed up graphically by Cloudsely-Thompson (1951). He reported that 'removal or painting over the antennae induces a state of depression which persists until death'.

The antennae are covered in a variety of sensory structures which include gustatory and olfactory receptors. Haacker (1974) and Carey and Bull (1986) have shown that male millipedes do not mate successfully with females if their antennae are removed. This is presumably because they can not detect the stimulatory pheromones produced by the female.

The antennae of all millipedes have eight sections called articles. The apical section is usually small and bears four characteristic cone-shaped sensilla (Fig. 6.7). The antennae have remarkable powers of regeneration (Nguyen Duy-Jacquemin 1972*a*). In *Polydesmus angustus*, the whole antennae will regenerate if it is amputated above the third article. However, it will only regenerate to a juvenile condition if amputated below the second article (Petit 1974*a*).

Probably the most in-depth studies on the sensory organs of millipedes have been those of Monique Nguyen Duy-Jacquemin on the antennae. Her results have been reported in a beautifully-illustrated series of papers published over the last two decades (Nguyen Duy-Jacquemin 1972*b*, 1981, 1982, 1983, 1985*a,b*, 1988, 1989, 1990). These have described the structure and function of the different types of sensilla. In julids, there are five types of sensory organs on the antennae (Fig. 6.8). The most prominent are the four apical cones (Figs. 6.9, 6.10, 6.11). Mechano- and chemoreceptors occur which have a similar internal structure to those of insects. Thermo- and hygroreceptors are probably present also. However, the function of the basiconic sensilla without pores is still not clear and more work is required before we have a full understanding of the stimuli to which the antennae of millipedes respond.

6.3 Neurosecretion

6.3.1 Introduction

A neurohaemal organ is a storage and release site in the nervous system that contains neurosecretory axons (Sahli 1979, 1985*a*). In comparison to insects, little is known about neurosecretion in millipedes (Sahli 1990). There is a pressing need for laboratory studies that relate neurosecretory activity to specific environmental, behavioural, and physiological factors. Several activities of millipedes such as moulting (Nair 1980), the reproduc-

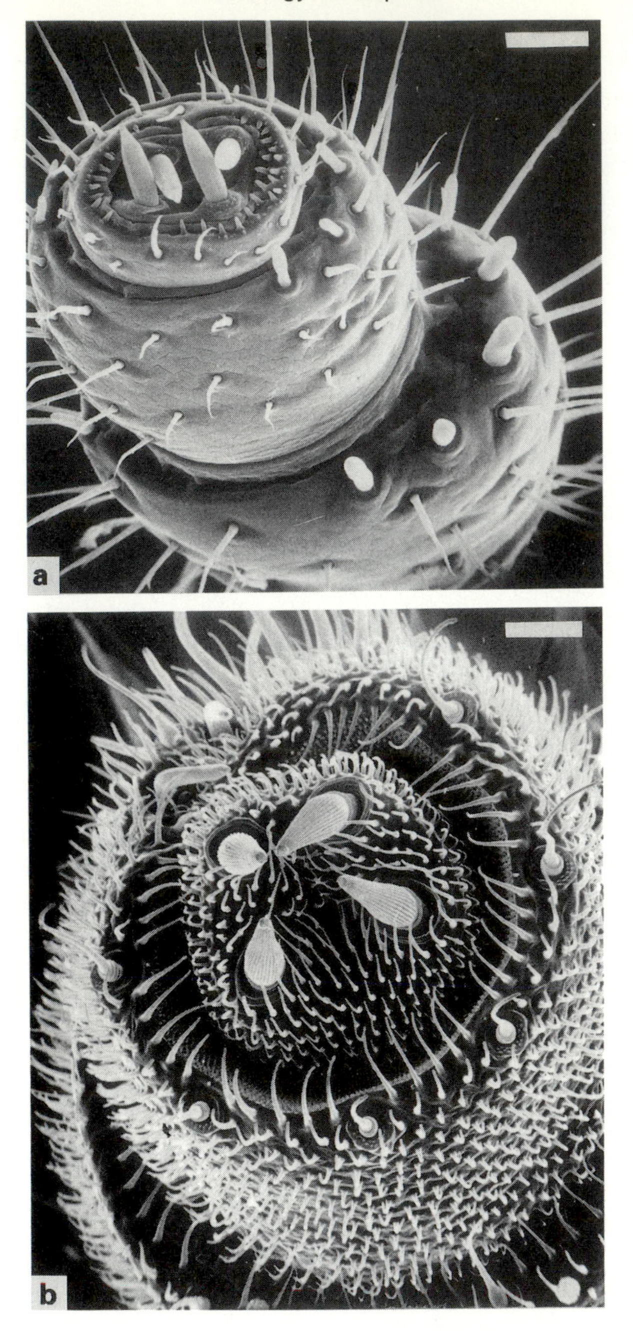
a
b

Fig. 6.7 Scanning electron micrographs of the terminal articles of the antennae of two species of millipede showing the four conical sensilla at the tip. (a) *Cylindroiulus punctatus* (see also Fig. 6.8). Scale bar 25 μm. Reproduced from Nguyen Duy-Jacquemin (1985*b*) by kind permission of the author and Artis Bibliotheek, Amsterdam. (b) *Polyxenus lagurus*. Scale bar = 5 μm. Reproduced from Eisenbeis and Wichard (1987) by kind permission of the authors and Springer-Verlag.

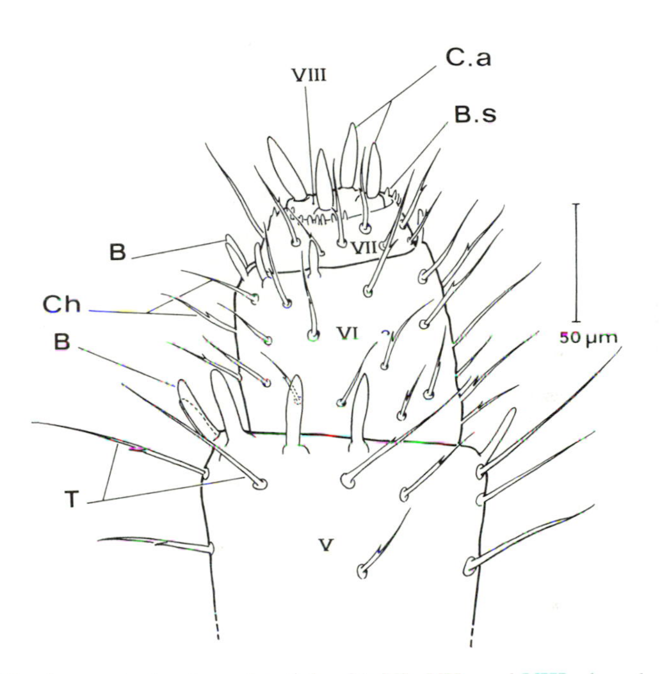

Fig. 6.8 The four apical antennal articles V, VI, VII, and VIII, dorsal side, of an adult *Cylindroiulus punctatus* (see also Fig. 6.7a) showing the five types of sensory organs of julid millipedes. B, sensilla basiconica; B.s, sensilla basiconica spiniformis; C.a, apical cones; Ch, sensilla chaetica; T, sensilla trichoidea. Reproduced from Nguyen Duy-Jacquemin (1985*a*) by kind permission of the author and S. A. Masson (Annales des Sciences Naturelles).

tive cycle (Rosenberg and Warburg 1982; Warburg and Rosenberg 1983), and some metabolic processes such as regulation of blood glucose levels (Satyam and Ramamurthi 1978) are almost certainly under neurosecretory control. However, there is almost no experimental evidence in support of this.

There is an extensive literature on the structure (as opposed to the functions) of neurohaemal organs in millipedes. Sahli and co-workers have been most active in this area since the first descriptions appeared some 30 years ago (Sahli 1958*a*, 1961*a*, 1962). Other significant studies have been conducted by Juberthie-Jupeau (1967*a,b*, 1973) and Nair (1973, 1980). The

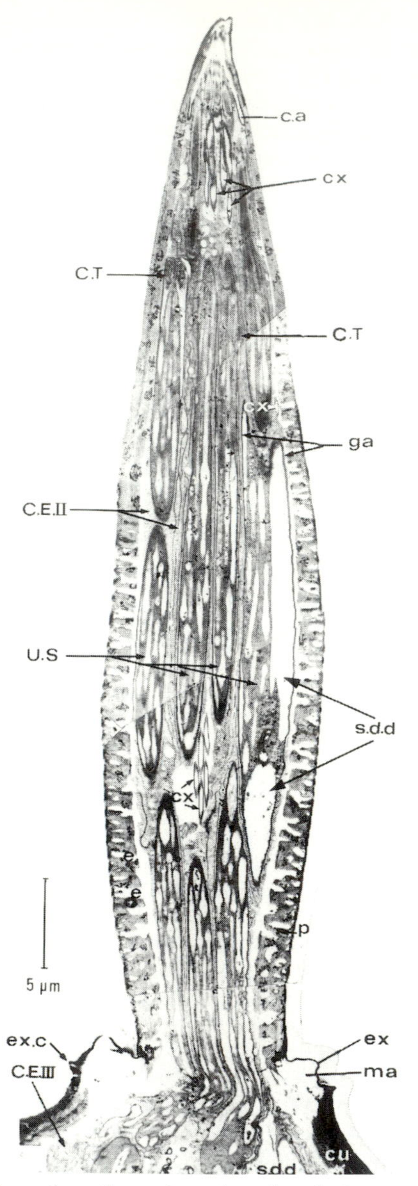

Fig. 6.9 Longitudinal section through a dorsal apical cone of the antennae of *Typhloblaniulus* (=*Blaniulus*) lorifer (the section does not pass through the apical pore which can not be seen). c.a, apical circular invagination; C.E.II, C.E.III, sheath cells II and III; C.T, tubular bodies; cu, cuticle; cx, ducts; e, lacunae in the wall (p); ex, exocuticle; ex.c, cuticular outgrowths of the insertion prominence (ma); ga, cuticular sheath; s.d.d, expanded distal segment; U.S, sensory units. Reproduced from Nguyen Duy-Jacquemin (1985*a*) by kind permission of the author and S. A. Masson (Annales des Sciences Naturelles).

literature has been difficult to interpret due to a lack of a consistent terminology for the different organs. However, a number of reviews have appeared in recent years which have removed some of the confusion (Juberthie-Jupeau 1983; Sahli 1985*a*; Joly and Descamps 1987, 1988).

6.3.2 Neurohaemal organs

Six neurohaemal organs have been described in the brain of millipedes, although not all species possess all six types (Juberthie and Juberthie-Jupeau 1974). The typical arrangement of those in Julida is shown in Fig. 6.12. However, it should be borne in mind that this is very much a preliminary diagram and that much remains to be discovered about neurosecretion in millipedes.

6.3.2.1 Gabe's organ

A pair of Gabe's organs are probably present in all diplopods. Also known as the cerebral glands, they were discovered by Gabe (1954). The organ is composed of axon terminals and glial cells and contains no intrinsic secretory cells. Gabe's organs are connected to the brain via neurosecretory axons which release their secretions in globulus I (Fig. 6.12). The structure of these organs has been described in a number of species from several orders (Nguyen Duy-Jacquemin 1974); Petit and Sahli 1977; Prabhu 1961; Sahli 1977*b*; Sahli and Petit 1972; Seifert 1971). Gabe's organ is probably a primitive structure. It may be analogous to the neurohaemal organ of Collembola, Diplura, and some Symphyla and Thysanura (Sahli 1974), and could be involved in the control of moulting.

6.3.2.2 Paraoesophageal bodies

Paraoesophageal bodies are oval in shape and are located ventro-laterally to the brain. The organ contains intrinsic secretory cells and may be equivalent to the corpus cardiacum of insects (Sahli 1974). Their structure has been described by Juberthie-Jupeau (1973), Petit and Sahli (1975), Sahli (1977*a*), and Sahli and Petit (1973). Their secretory products are released partially into blood sinuses and probably also into globulus I in the brain, which receives secretions from Gabe's organs as well (Petit and Sahli 1975).

6.3.2.3 Connective bodies

Connective bodies are present in *Oxidus (Orthomorpha) gracilis*, but have not been found in julid millipedes (Sahli and Petit 1974). They are cephalic neurohaemal organs which are connected to a neurosecretory tract that runs from the brain along the central nerve cord. Sahli (1974) considered them to be vestiges of a primitive metameric arrangement of paired neurohaemal organs. They may be analogous to the perisympathetic organs of

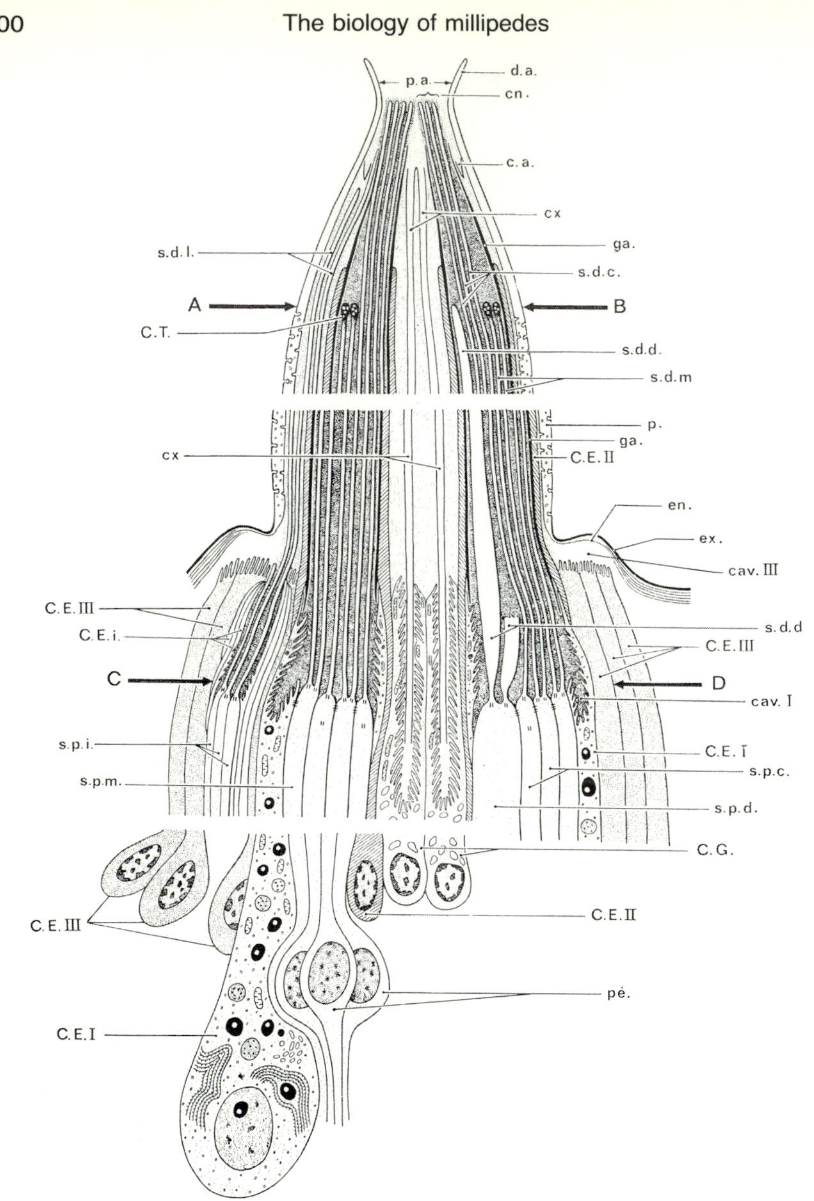

Fig. 6.10 Diagram of a longitudinal section through a dorsal apical cone of the antennae of *Typhloblaniulus* (=*Blaniulus*) *lorifer* through two sensory units. The apical region of the right unit includes the neuron with the expanded dendrite. c., basal body; c.a., apical circular invagination; cav I and III, inner and outer receptor lymph cavities; C.E.I, and C.E.II, C.E.III, C.E.IV and C.E.i., sheath cells I, II, III, IV and of the independant neuron group; C.G., glandular cells; cn., cord formed from the chemoreceptive distal segments, their sheath and the electron-dense substance; C.T., tubular bodies; cx, ducts; d, desmosome; d.a., apical finger-

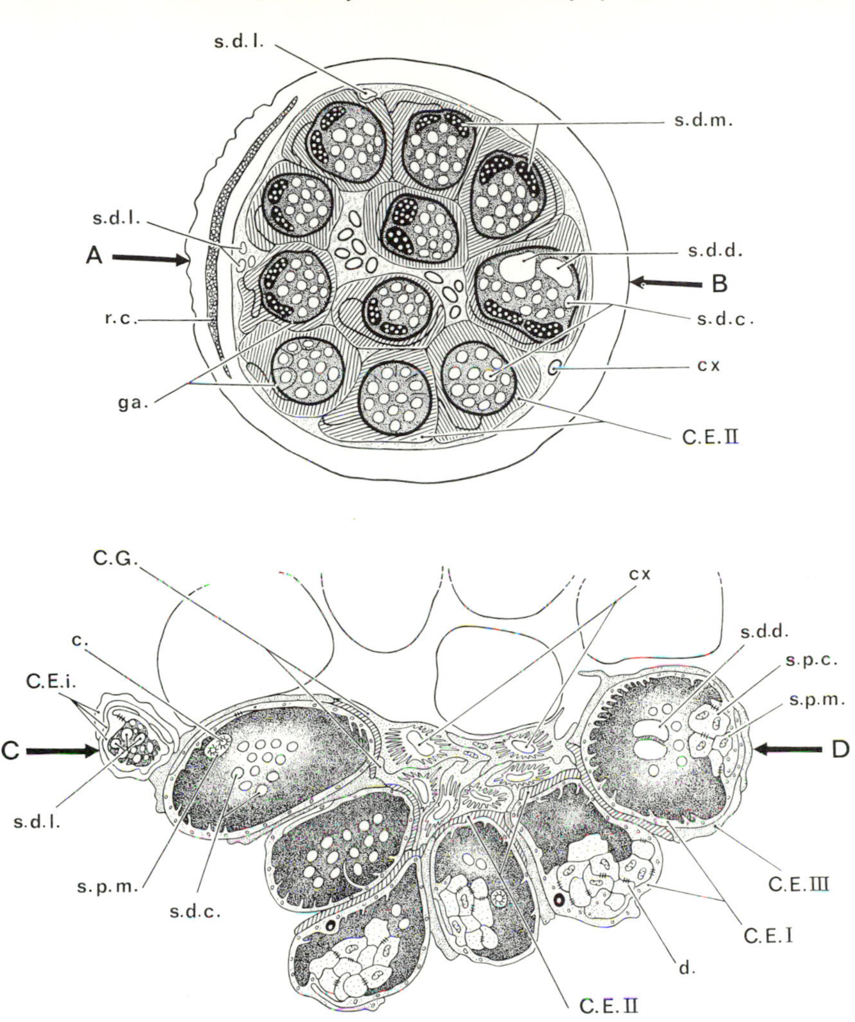

Fig. 6.11 Transverse sections of a dorsal apical cone of the antennae of *Typhloblaniulus* (=*Blaniulus*) *lorifer* through planes A–B and C–D. For legends and source see Fig. 6.10.

---

like structures; en., laminated endocuticle; ex., exocuticle; ga., cuticular sheath; p., cone wall; p.a., apical pore, pé., perikarya; r.c, network of ductules; s.d.c., s.d.d., s.d.l., and s.d.m., chemoreceptive, expanded, free, and mechanoreceptive distal segments; s.p.c., s.p.d., s.p.i., s.p.m., chemoreceptive, expanded, of the independent neurons and mechanoreceptive proximal segments. A, B, and C, D, planes of transverse section shown in Fig. 6.11. Reproduced from Nguyen Duy-Jacquemin (1985*a*) by kind permission of the author and S. A. Masson (Annales des Sciences Naturelles).

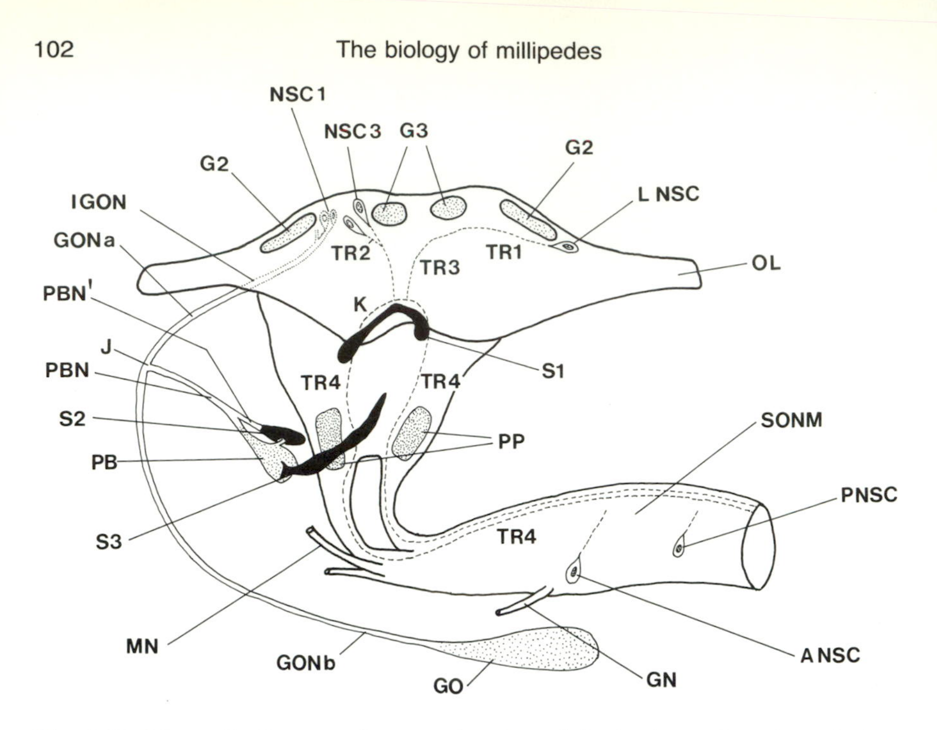

Fig. 6.12 Schematic diagram of a posterior view of the brain, periosophageal ring, and suboesophagel nerve mass of a julid millipede. Only the left neurohaemal complex has been drawn. ANSC, anterior neurosecretory centre of the suboesophageal nerve mass; G2 and G3, globulus 2 and 3; GN, gnathocilarium nerve; GO, Gabe's organ; GONa and GONb, a and b parts of the Gabe's organ nerve; IGON, intracerebral course of the Gabe's organ nerve; J, level of junction of the Gabe's organ and paraoesophageal body nerves; K, K commissure; MN, mandibular nerve; NSC1, neurosecretory centre of Gabe's organ nerve (situated on the anterior side); NSC3, neurosecretory centre of globulus 3; OL, optic lobe; PB, paraoesophageal body; PBN, paraoesophageal body nerve; PBN', nerve joining the blood sinus formation S2 to the paraoesophageal body nerve; PNSC, posterior neurosecretory centre of the suboesophageal nerve mass; PP, paracommissural plates; S1–3, blood sinus formations S1–3; SONM, suboesophageal nerve mass; TR1–4, pathways of neurosecretory products. Redrawn from Joly and Descamps (1988) by kind permission of the authors and Alan R. Liss Incorporated. Most of the findings on which this diagram is based were performed by Sahli and co-workers.

pterygote insects. Nair (1974—quoted in Joly and Descamps 1988) considered that they may control oocyte growth.

6.3.2.4 Perioesophageal blood sinus formations

These structures may be clearly separate from, or associated closely with, the paraoesophageal bodies. Their structure has been described in julids by Sahli and Petit (1979). The secretory products are released into the blood sinus located in the anterior part of the cephalic aorta.

6.3.2.5 Paracommissural plates

Paracommissural plates are not clearly separate from the brain. They consist of a pair of plates which lie near the transverse commissure of the circumoesophageal connectives (Sahli and Petit 1975). Paracommissural plates are composed of glial cells, tracheae, at least two distinct types of axon terminals, and a neural lamella facing a blood sinus into which secretions can be released. The axons of the paracommissural plates originate in tritocerebral neurosecretory cells lying in the ganglions of the transverse commissure (Petit and Sahli 1978*a*).

6.3.2.6 Perioesophageal gland

This gland has been found only in *Polyxenus lagurus* and is located dorsally between the pharynx and the brain. El-Hifnawi and Seifert (1972) considered it to be an ecdysial gland.

7

Gametogenesis and fertilization

7.1 Introduction

Millipedes have separate sexes. Secondary sexual structures have evolved to a high degree to ensure successful transfer of sperm from males to females. In this chapter, all aspects of reproduction will be covered, from the structure of reproductive organs and gametes, through to mating behaviour, egg laying, and early embryological development. Post-hatching development is covered in Chapter 8.

7.2 Reproductive organs

7.2.1 Introduction

The reproductive organs of both sexes are situated between the gut and the ventral nerve cord. They open either on ring 3 or just behind the second pair of legs (Blower 1985).

7.2.2 Ovaries

The ovaries of female millipedes are paired. In most orders they are housed in a common median ovitube which bifurcates anteriorly into short oviducts. These open into the vulvae behind the second pair of legs. In julids, the paired ovaries within the ovitube run from about ring 15 to the last podous ring. In a ripe female, the eggs are packed tightly within the tube which occupies half or more of the ring volume (Fig. 7.1). Eggs are passed anteriorly in a single row for laying.

In their detailed study on *Glomeris marginata*, Heath *et al.* (1974) showed that the number of eggs in the ovaries exhibited a pronounced seasonal cycle (Fig. 7.2). Mature eggs were about 1 mm in diameter (in animals *c.* 12 mm in length). This species clearly spends the autumn and winter building up eggs which are released in the spring and early summer.

The ultrastructure of the ovaries of *Ophyiulus pilosus* and *Polyxenus lagurus* has been studied recently by Kubrakiewicz (1991*a,b*).

7.2.3 Testes

The males of most species have a pair of tubular testes that run the whole length of the body. There is a gradient in maturity of sperm from posterior

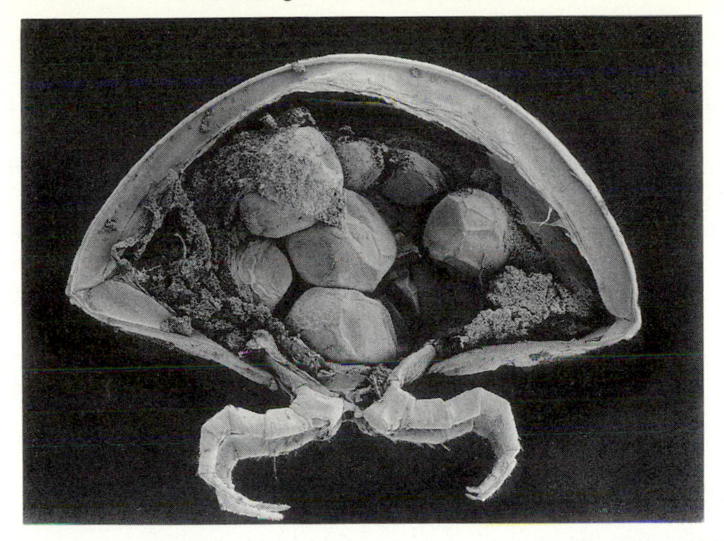

Fig. 7.1 Scanning electron micrograph of a cross section through a mid-body segment of a female *Polyzonium germanicum*. The segment is 2 mm in diameter. Note the large eggs which occupy a considerable proportion of the body cavity. Reproduced from Enghoff (1990) by kind permission of the author and E. J. Brill, Leiden.

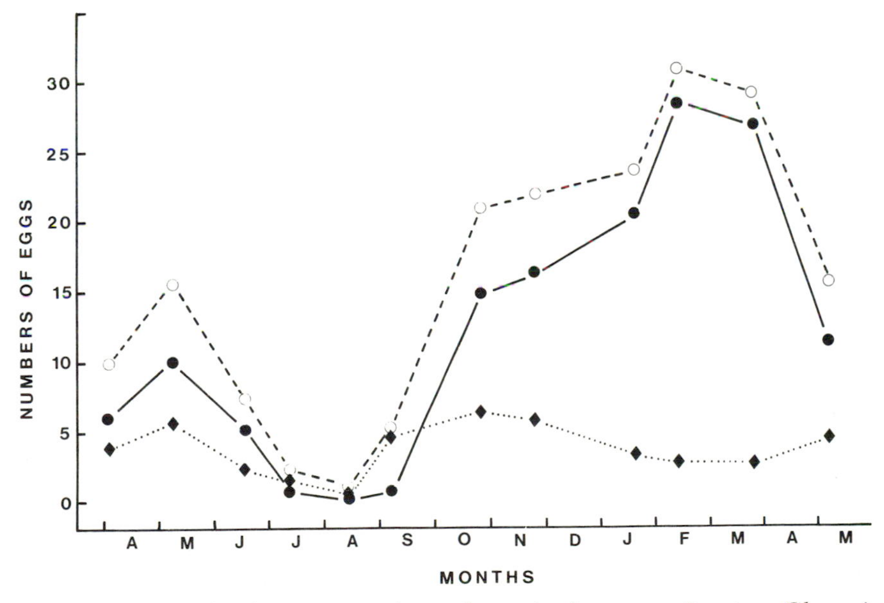

Fig. 7.2 Changes in the mean numbers of eggs in the ovary of mature *Glomeris marginata* from April 1963 to May 1964. The symbols represent mature eggs (●), developing eggs (◆), and total large eggs (○). Redrawn from Heath *et al.* (1974) by kind permission of the Zoological Society of London.

to anterior with the most mature sperm at the anterior end (Bessiere 1948; Petit 1974*b*). Each testis is surrounded by muscle fibres (Camatini and Castellani 1978) and opens separately on the coxae of the second pair of legs, or through a bilobed penis just behind these legs.

In most species, the testes are connected by transverse connections like the rungs of a ladder (Fontanetti 1988; Kanaka and Chowdaiah 1974). In Glomerida, there is a single median testes tube with a row of pouch-like follicles on either side (Blower 1985). Demange (1988) considered the paired condition to be primitive and the unpaired condition derived. This tube bifurcates anteriorly into a pair of vasa deferentia which open on the coxae of the second pair of legs. West (1953) has given an extremely detailed description of the development of the male reproductive system of *Scytonotus virginicus*.

7.3 Gametes

7.3.1 Spermatozoa

The spermatozoa of millipedes have been extremely well studied. There are several papers that describe their ultrastructure in great detail. Species and genera covered include *Glomeris marginata* (Boissin *et al.* 1972), *Polyxenus lagurus* (Baccetti *et al.* 1974), *Spirobolus* sp. (Reger 1971; Reger and Fitzgerald 1979), *Polydesmus* sp. (Reger and Cooper 1968), and *Pachyiulus enologus* (Camatini and Franchi 1978). This work, and that of others, has been summarized in several excellent reviews (Bacetti and Dallai 1978; Manier and Boissin 1978; Jamieson 1987).

Animals with sophisticated reproductive behaviour such as millipedes (see Section 7.4.6) tend to have immotile sperm. Indeed, there are no known examples in Diplopoda of flagellated sperm (Fig. 7.3). In julids, the acrosome reaction involves the extrusion of a long filament (Baccetti *et al.* 1977). This has been misinterpreted as a flagellum by some authors.

During the evolution of millipede sperm, a primary divergence appears to have occurred between the penicillate (bristly) and chilognath millipedes. Bristly millipedes are distinguished by the ribbon-like sperm that retain a monolayered acrosome and mitochondria while the chilognaths have what may be termed 'biscuit-shaped' sperm (Baccetti *et al.* 1979). The bristly millipedes differ further in that the shape of the sperm changes from barrel-shaped in the male ducts to ribbon-shaped in the female ducts. This does not occur in chilognaths.

Jamieson (1987) has shown that the ultrastructure of the sperm of the different orders of millipedes is quite distinct. Indeed, they provide strong supporting evidence for the classification of the Diplopoda which has been derived from more traditional morphological analysis.

The sperm are 'packaged' into a spermatophore which is passed to the female during mating. The package is produced by coating a collection of

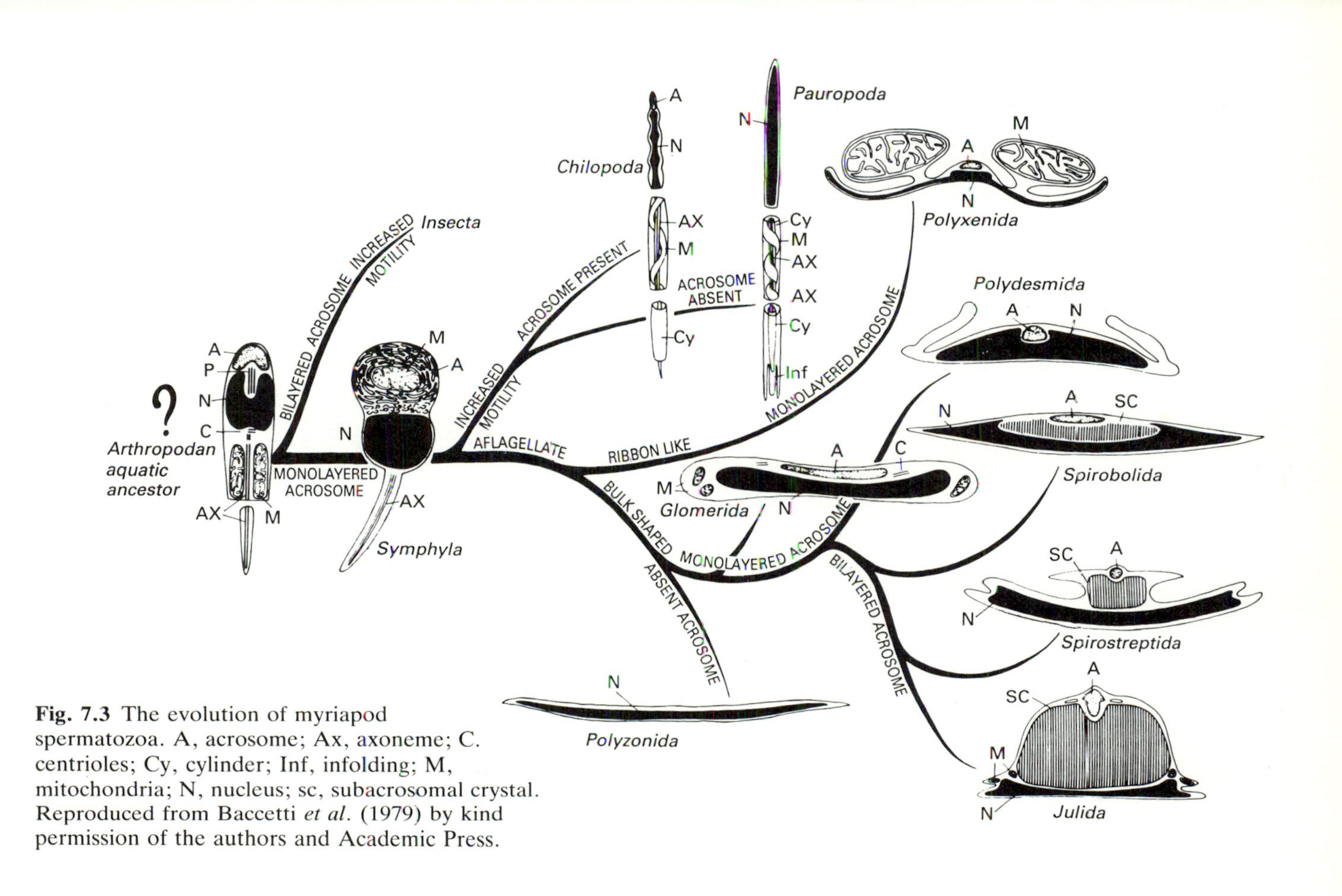

Fig. 7.3 The evolution of myriapod spermatozoa. A, acrosome; Ax, axoneme; C. centrioles; Cy, cylinder; Inf, infolding; M, mitochondria; N, nucleus; sc, subacrosomal crystal. Reproduced from Baccetti *et al*. (1979) by kind permission of the authors and Academic Press.

sperm with secretions from accessory glands as they pass down the vas deferens (Demange 1988; West 1953). Very little is known about the formation of the spermatophore and this would clearly repay further study.

7.3.2 Eggs

The eggs of millipedes are quite yolky. They have to provide the developing embryo with all its nourishment until after the second moult, when the animal begins to feed. As well as sufficient lipid, carbohydrate, and protein, the eggs are provided with mineral elements such as calcium which is present in granules (Crane and Cowden 1968). The granules may be dissolved and the calcium used for the preliminary calcification of the exoskeleton before the element is obtained for this purpose from the food (Hubert 1975, 1977, 1979*a,b*; Petit 1970). Alternatively, the calcium granules may be the end product of a storage-detoxification system for metals (Hopkin 1989; Kubraktewicz 1989).

The development of ovocytes in four species of millipedes has been described in detail by Crane and Cowden (1968). The females may also supply the eggs with cyclopropane fatty acids. These substances probably provide resistance to desiccation by forming a relatively impermeable outer layer (Horst and Oudejans 1973, 1978; Horst *et al*, 1972, 1973; Oudejans 1972*a,b*; Oudejans and Horst 1978; Oudejans *et al*. 1976).

7.4 Mating

7.4.1 Introduction

It is important to recognize the difference between insemination and fertilization in millipedes. The process of insemination occurs during mating when the male supplies the female with a packet of sperm in a spermatophore. She takes the sperm into blind-ending sacs called the spermathecae or seminal receptacles. The sperm are stored in these sacs and the eggs are not fertilized until just before laying when sperm are allowed onto the eggs as they leave the body at oviposition.

In *Polyxenus lagurus*, the spermatophore of the male is deposited on a small silken web (Fig. 7.4). The female is attracted to this and draws it into her genital opening. In contrast, all chilognath millipedes have direct sperm transfer via complex secondary sexual organs called gonopods. These are analogous to the lock-and-key system of the palps and epigynes of spiders. The males flex their bodies, take up the spermatophore in their gonopods from their genital opening, and use these intromittent organs to insert the package of sperm into the genital opening of the female.

There is ample evidence of parthenogenicity in millipedes (Brookes 1974; Enghoff 1976*a,b*; Fussey and Varndell 1980; Nguyen-Duy Jacquemin and Goyffon 1977) although the number of species in which this has been

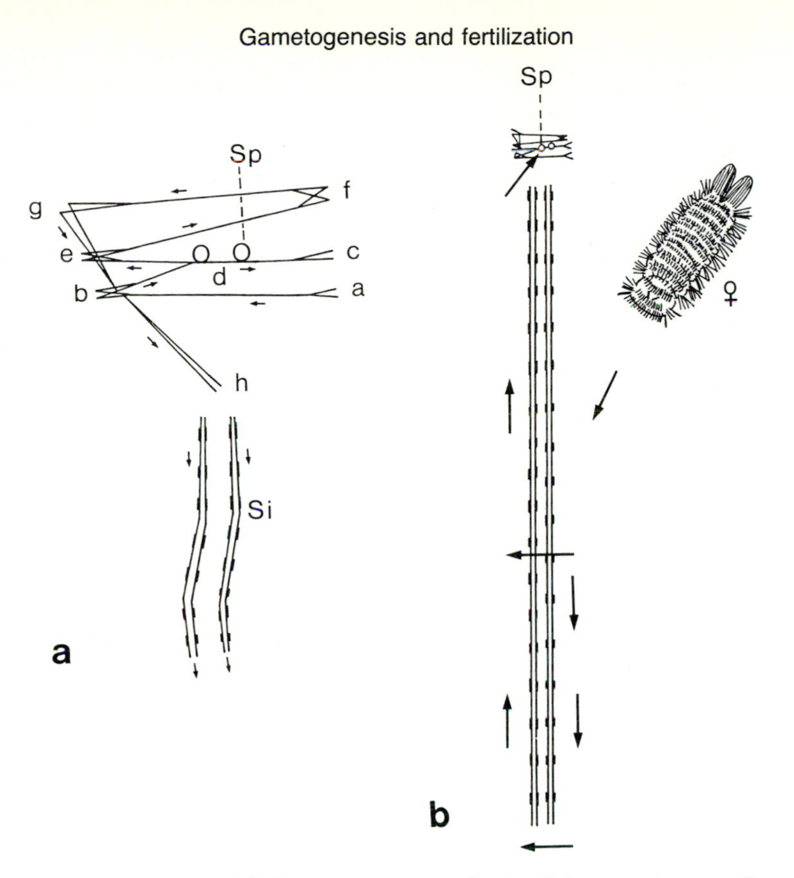

Fig. 7.4 The method of indirect sperm transfer in *Polyxenus lagurus* (2 mm in length). (a) Two droplets of sperm (Sp) are deposited on a web of threads (secreted by penis glands) spun with zigzag movements by the male. In addition, a trail of signal threads (Si) of approximately 1.5 cm in length is secreted by the glands on the eighth and ninth pair of legs. (b) A mature female recognizes the signal threads with her antennae and either crosses or walks round to the other side of the signal trail which leads her to the sperm web. When she finds the mesh of threads, she takes the sperm droplets into her genital valves. Reproduced from Eisenbeis and Wichard (1987) by kind permission of the authors and Springer-Verlag. The diagram is based on the work of Schömann (1956).

described is low (only about 20; Enghoff 1978*a*). This topic is described in more detail in Section 8.5.5.

7.4.2 Gonopores

Spermatophores in males, and eggs in females, emerge from gonopores associated with the second pair of legs. Male Penicillata, Pentazonia, some Colobognatha, Callipodida, Chordeumatida, and Polydesmida have gonopores that open on the second coxae, or through penises on them. In contrast, Stemmiulida, Juliformia, and some Colobognatha have gonopores (on paired or secondarily unpaired penises) posterior to the second pair of legs (Enghoff 1990).

In female diplopods, only the Penicillata and Glomeridesmida have gonopores which open on the second coxae. All other female millipedes have gonopores that open posterior to the second coxae.

7.4.3 Gonopods

Millipedes possess a number of secondary sexual characters which are designed to improve the efficiency with which spermatophores are transferred from males to females.

Each oviduct of female millipedes opens separately into organs called vulvae which are contained within separate sacs in the lumen of the ring posterior to the second pair of legs (Fig. 7.5). These structures are everted during copulation. Each vulva consists of a bursa (rather like the hinged shells of a bivalve mollusc) with the opening directed anteriorly and covered by an operculum (Blower 1985). The detailed structure of the vulvae is unique to each species and may be of considerable importance in identification (Kurnik 1988; Kurnik and Thaler 1985).

In the body of the vulva there is an apodemic tube which ends in one or two blind-ending ampullae. These are spermathecae, receptacles for semen. It is interesting to note that in parthenogenetic races of millipedes,

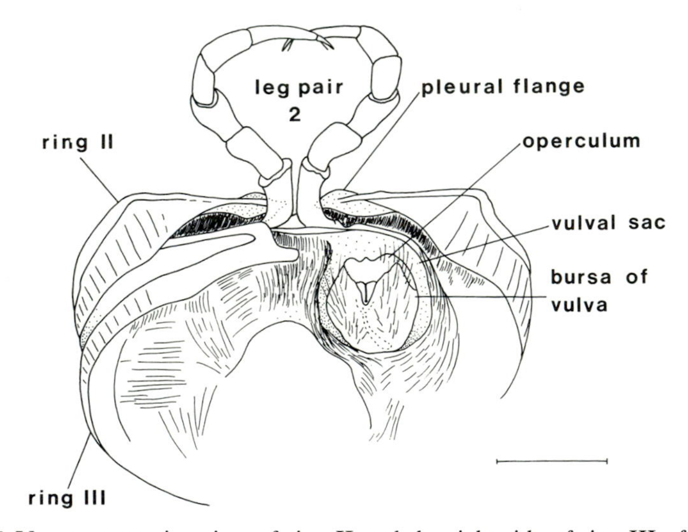

Fig. 7.5 Ventro-posterior view of ring II and the right side of ring III of a female *Tachypodoiulus niger* with the second pair of legs and the vulvae posterior to them. On the animal's left, the pleurite of ring III has been removed and the vulval sac trimmed to reveal the vulva within. Scale bar = 0.5 mm. Redrawn from Blower (1985) by kind permission of the author, the Linnean Society of London and Academic Press.

females often have a considerably reduced spermathecae (Enghoff 1976*b*, 1979*b*).

In the Penicillata (bristly millipedes), there are no other obvious secondary sexual structures. In males of Glomerida, the most posterior three pairs of legs in males are modified as gonopods, the last pair being termed telopods. In almost all other male millipedes, one or both pairs of limbs on the seventh ring (limb-pairs 8 and 9) are modified to form the structures by which sperm are introduced into the female (Fig. 7.6). The fine structure of the gonopods may be the single most important character for specific identification.

The development of male gonopods has been studied in considerable detail in several orders of millipedes (Blower and Gabbutt 1964; Petit 1973, 1976; West 1953). Berns (1968) showed that in *Narceus annularis* (Spirobolida), the male gonopods develop from rudiments of both pairs of

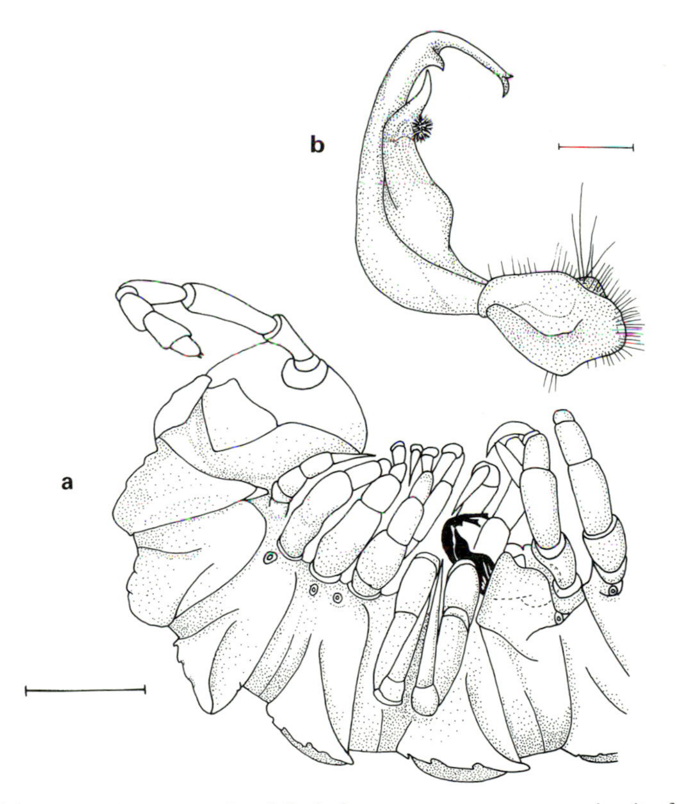

Fig. 7.6 (a) The male gonopods of *Polydesmus angustus* as seen *in situ* from the right side. Legs 6 and 7 have been displaced to reveal the gonopods (shown in black). Scale bar = 1 mm. (b) Enlarged view of the right gonopods. Scale bar = 0.2 mm. Redrawn from Blower (1985) by kind permission of the author, the Linnean Society of London and Academic Press.

legs on the seventh body ring. They are first evident as small bumps on the ventral side of the seventh ring following the moult to the fifth instar, replacing the functional walking legs of the seventh ring of earlier instars. These protrusions go through progressive morphological changes in each instar until they attain the adult form. It is clear from Berns's work that in spirobolid millipedes the gonopods do not develop as a gradual modification of functional walking legs. They pass through a progressive growth and differentiation of their own. Similar findings were made by Petit (1973) on male *Polydesmus angustus* in which the eighth pair of walking legs develop into gonopod buds at the end of the third instar.

The gonopods can be incredibly complex and may not resemble legs in the slightest. The structure consists of a pair of gonopods, one for each lateral half of the body, which are mirror images of each other. They are sclerotized and are much harder than the vulvae of the females. A whole vocabulary has developed to describe various folds and projections on them.

Several specimens have been discovered in the wild which possess gonopods that are either of unusual structure or are on the 'wrong' ring. Shelley (1977) described a number of these in the Xystodesmidae (Polydesmida). One particular specimen had an appendage in the normal reproductive position which had both leg-like and gonopod-like segments. Thus, millipedes possess information to produce an appendage on ring 7 which can either be a leg or a gonopod. Which of these differentiates must depend on as yet unidentified chemical signals. In the specimen shown in Fig. 7.7, these signals have clearly become confused!

Petit (1973) described a number of experiments in which he amputated the eighth legs of stadium III *Polydesmus angustus* at different points. The nature of the structure which developed depended on the point of amputation. In general, the more distal the amputation, the more typical the

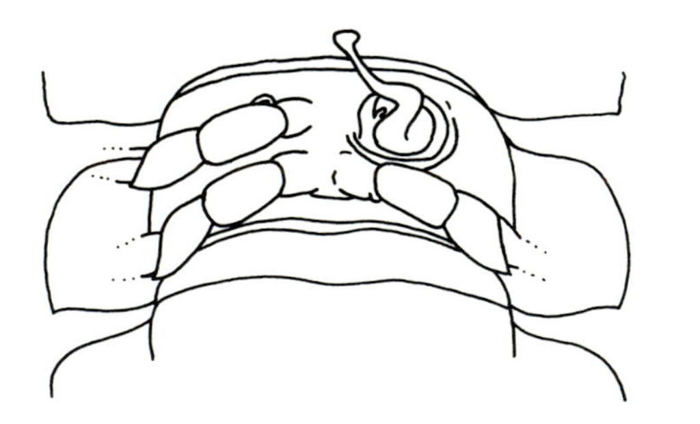

Fig. 7.7 Ventral view of the seventh segment of the male holotype of *Fontaria rubromarginata*. The specimen exhibits *hysterotely*, having a gonopod on the left side only. A leg has developed instead of a gonopod on the right side. Redrawn from Shelley (1977) by kind permission of the author and the National Research Council of Canada.

structure of the gonopod which subsequently developed. Thus damage to the limb itself can influence subsequent development and at least some of the abnormalities noted in millipedes could be due to such damage in earlier stadia (Balazuc and Schubart 1962).

7.4.4 Other secondary sexual characteristics

In addition to the gonopods, several millipedes have developed other secondary sexual characters. These have been reviewed for Julida by Petit and Sahli (1978*b*). They recognized five main structures:

(1) glands that open on the coxae of the second pair of legs, which are sometimes swollen;

(2) an 'inflated' gnathocilarium;

(3) first pair of legs modified into a 'crotchet-like' structure;

(4) ventrally produced flanges on the cardo and stipes of the mandibles; and

(5) adhesive pads on the legs which help the male hold on to the female during mating.

All these structures in Julida, and similar ones on other orders, are involved in mating (see Krabbe (1979) for a discussion of their role in spirostreptids).

7.4.5 Pheromones

There is clear evidence that millipedes may be attracted to each other by chemical signals. In male *Glomeris marginata*, a pheromone is produced from the post-gonopodial gland which opens between the 19th (last) pair of legs and the anus (Juberthie-Jupeau 1976). This gland is of ectodermal origin and arises from an invagination of the integument (Juberthie-Jupeau and Tabacaru 1968). However, it is not clear over what distance these signals can operate. In *Ommatoiulus moreleti*, experiments in mazes with controlled air flow have shown that males are not attracted to the smell of females (Carey and Bull 1986). Thus, secretions from the glands may only act over very short distances. This is an area requiring more research. For example, it may be important in pest control if communication between the sexes can be disrupted.

7.4.6 Mating behaviour

In *Polyxenus lagurus*, the male spins a mesh of threads in a zigzag pattern in a small crevice, then fixes two droplets of sperm onto one of the stretched threads (Fig. 7.4). Subsequently, he stretches a conspicuous, double thread (*c*. 15 mm in length) vertically downwards. A passing female, perceiving the thick double thread, follows it and is led directly to the two sperm droplets which she then takes into the valves on her second pair of legs. Thus sperm transfer in Penicillata does not require direct contact between male and female (Eisenbeis and Wichard 1987).

The mating behaviour of chilognath millipedes is quite elaborate (Figs. 7.8, 7.9). Haacker and Fuchs (1970) made detailed observations of this in *Cylindroiulus punctatus* which they considered to be typical for most Julida. The females are relatively passive during mating; the males take the active role.

In *Cylindroiulus punctatus*, the male runs up to the back of the female and clings to her using the pads on his legs. He then moves his front end

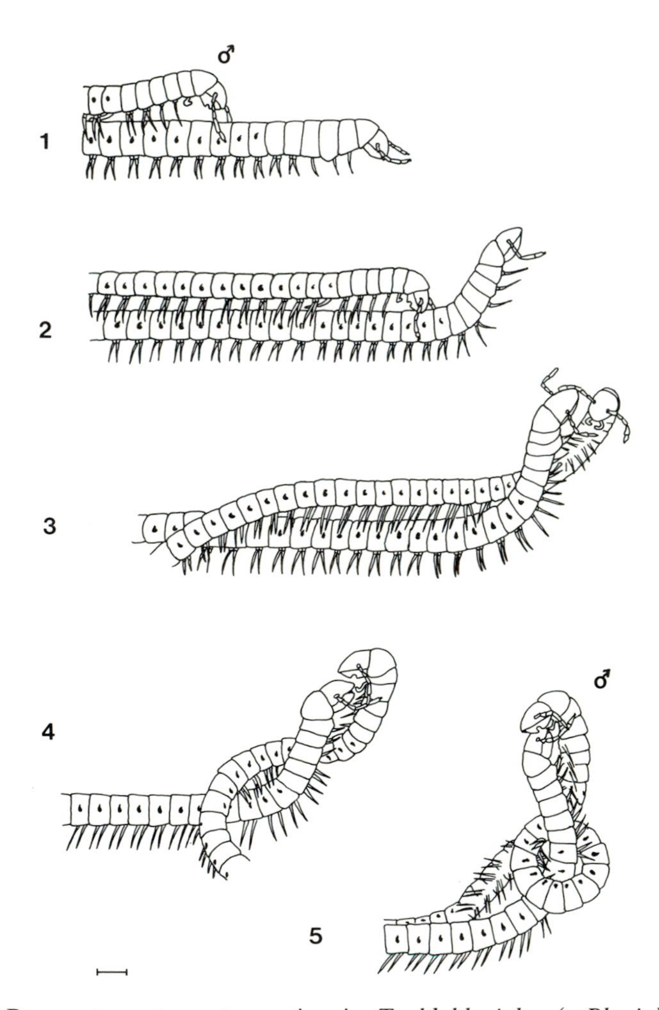

Fig. 7.8 Preparatory stages to mating in *Typhloblaniulus* (=*Blaniulus*) *lorifer consoranensis*. Scale bar = 0.5 mm. Redrawn from Mauriès (1969) by kind permission of the author and the Director of Laboratoire Souterrain, Saint-Girons.

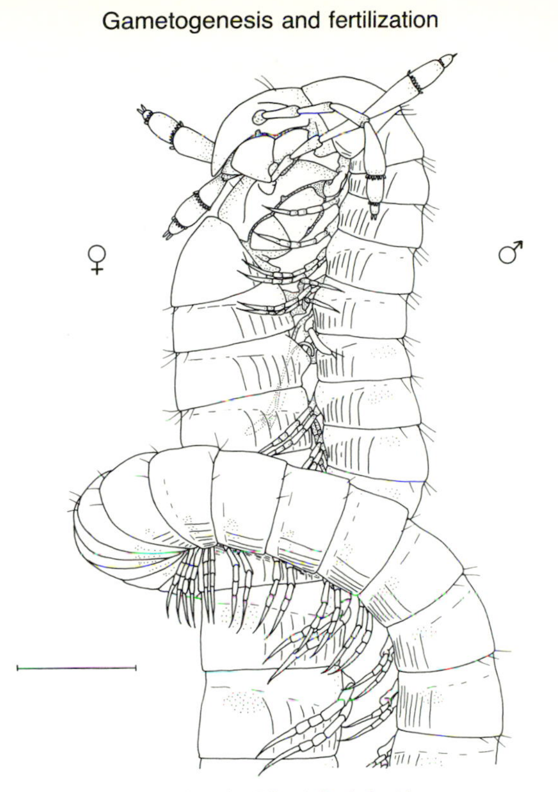

Fig. 7.9 Mating in *Typhloblaniulus* (=*Blaniulus*) *lorifer consoranensis*. The gonopods of the male are shown inserted into the genital opening of the female (dotted lines). Scale bar = 0.5 mm. Redrawn from Mauriès (1969) by kind permission of the author and the Director of Laboratoire Souterrain, Saint-Girons.

ventrally so that the two sexes face each other (ventral surfaces together). The head of the female is fixed in position by the first pair of legs of the male. The vulvae by leg-pair 2, are pulled out by the male gonopods on ring 7. If the female is not ready to mate she may withdraw her vulvae. The male inserts the opisthomerites (posterior part of the gonopods—free from sperm at this stage) and conducts rhythmic movements, presumably to prepare for sperm transfer. He then flexes his body and takes sperm from the erected genital papilla into the opisthomerites, which are held forward. The vulvae are pulled out again, the sperm-laden opisthomerites are inserted into the vulvae, and insemination takes place.

Mating may last from a few minutes to several hours and may be repeated several times (Bercovitz and Warburg 1988). In *Alloporus uncinatus*, the duration of copulation is prolonged in the presence of other males, providing support for the copulatory guarding hypothesis (Telford and Dangerfield 1990).

There are numerous variations and elaborations on this basic theme. In *Julus scandinavius*, which Haaker (1969) considered to be the more advanced state, the male presents a secretion to the female prior to mating, which he produces from the glands on the coxae of the second pair of legs. The female licks the secretion while assuming the mating position, thus allowing the male to insert his charged gonopods. In *Archispirostreptus gigas*, the female holds tightly onto the prefemoral processes on the first pair of legs of the male (Krabbe 1979). Females of *Choneiulus subterraneus* have a unique modification of the second antennomere which the male hangs onto with his 'parrot-bill' shaped mandibles during mating (Enghoff 1984*b*).

Haaker (1974) has reviewed mating behaviour in millipedes. Activities include stridulation, drumming on the ground, and touching. *Loboglomeris pyrenaica* has a group of longitudinal ribs on the femur of the telopods and approximately 350 fine striations on the ventral margin of the pygidium. Stridulation is an intergral part of the sexual behaviour. The male holds one antenna and vulva of the female with his telopods then, with lateral movements of the posterior part of the body, rubs the telopods with the pygidium. The female probably senses the vibrations (Haacker 1969). Oscillograms of stridulation in closely-related species of Sphaerotheriida are quite distinct (Haacker 1974). In overcrowded conditions, the signals become confused and males attempt to mate with males. (Mukhopadhyaya and Saha 1981).

The female may also produce an attractive signal. If the antennae of *Ommatoiulus moreleti* are removed from a female then mating takes place as normal, but if they are removed from a male, he will not initiate courtship (Carey and Bull 1986). There must be some property of the female cuticle that the male recognizes with his antennae, which stimulates him into action.

In *Polydesmus inconstans*, females appear to be able to get enough sperm from one mating to fertilize all their eggs. In this species, mating takes place in the autumn and eggs are not laid until the following spring when the weather is warmer (Snider 1984*b*). Snider (1981*b*) considered that *Polydesmus inconstans* could probably get enough sperm from one mating for 7 to 8 ovipositions, producing over 1000 fertile eggs in a lifetime. In the laboratory, up to 22 ovipositions containing fertile eggs were recorded.

7.5 Genetics

Aside from parthenogenesis (Section 8.5.5), millipede genetics is a research area in which very little is known. White (1979) reviewed millipede cytogenetics. He concluded that all previous findings on diplopod chromosome numbers made before the development of modern microscopical and histological techniques should be treated with extreme caution. Hence we have no information on the distribution of heterochromatin, satellite DNA, nucleolar organizers, and other special features and markers.

Recent studies by Achar (1986, 1987) have included checklists of chromosome numbers of 60 species in which $2n$ ranges from 12 to 30. However, the caution expressed by White (1979) should be borne in mind. Clearly, more research is needed in this area.

7.6 Egg laying

Apart from male platydesmids (Hoffman 1982), there have been no reports of millipedes guarding their eggs. However, female millipedes take considerable care to protect their eggs by surrounding them with a resistant coat. Newport (1841) appears to have been the first researcher to notice this. He described female *Tachypodoiulus niger* (Julida) digging a hole in which the eggs were laid.

Polyxenus lagurus lays its eggs in a string stuck together like beads, and deposits them in a spiral to form a disc. The tail brush is pressed around the sticky eggs to encircle them in a ventilated protective sheath. This prevents direct contact between the eggs and the substrate (Eisenbeis and Wichard 1987).

Most chilognath millipedes produce a protective coat for their eggs by eating earth, then mixing the material with a secretion produced by glands in the rectum. As the faeces emerges, it is moulded by the anus to coat individual eggs (for example *Glomeris balcanica*, Iatrou and Stamou 1990*a*), or formed into a chamber in which the eggs are laid and subsequently sealed (Bhakat *et al.* 1989). Nematophorans such as *Craspedosoma simile* use silk for nest forming. The silk is produced from a gland that opens on a styloid process on the end of the telson (Wernitzsch 1910).

The chamber provides physical protection for the eggs and protects them from rapid fluctuations in humidity and temperature (Crawford *et al.* 1987). In deserts, protection from flooding is as important as protection from desiccation (Crawford and Matlack 1979). It has been suggested in the past that the walls of the chamber may possess antimicrobial properties but this has since been disproved (Eisner *et al.* 1970).

In *Narceus annularis*, egg capsules are taken into the rectum to be coated before deposition. The coating is made up of leaf material which the female chews into a pellet. Excess water is removed in the rectum. Shaw (1966) suggested that the coating may provide the first meal for newly-emerging young millipedes.

8

Development, moulting, and life histories

8.1 Embryonic development

Millipede embryology has been studied extensively by a number of workers and is a field with its fair share of contradictions. In centipedes, Minelli and Bortoletto (1988) have begun to relate studies on numbers of legs and segments to the rapidly developing work on *Drosophila* and 'segmentation genes'. However, in millipedes, we have to rely on more traditional histological methods at present.

The most comprehensive recent study was conducted by Dohle (1964) on *Glomeris marginata, Orthomorpha (Oxidus) gracilis, Polydesmus complanatus*, and *Polyxenus lagurus*. In a later publication (Dohle 1974*b*), he described the embryology of *Ommatoiulus sabulosus*. These papers should be consulted for more detailed descriptions of millipede embryology. The following account is based on Dohle's work.

In Diplopoda, the cleavage of the egg results in a blastoderm that completely surrounds the yolk. Eventually, the blastoderm cells on the future ventral side assume a columnar shape and become more crowded. This thickened part of the blastoderm will form the ectodermal germ band. The posterior end of the future germ band is marked by a plug of cells that pushes into the yolk. The mesoderm migrates into the yolk from the plug. The first indication of segmentation occurs at this stage when ectodermal cells and the mesoderm form into transverse bands which hollow out to form coelomic sacs.

Subsequent differentiation leads to the formation of rudiments of all major organs, appendage buds, and eventually, the pupoid stage (Fig. 8.1, 8.2).

8.2 Hatching and early development

Hatching is a slow process which was described in detail for '*Iulus terrestris*' (= *Tachypodoiulus niger*) by Newport (1841). The egg shell splits down the dorsal median line to about two-thirds of its length (Saudray 1953). The pupoid remains enclosed by membranes inside the split egg shell for about 12 hours. During this time, it remains attached by a funis inserted at the same position as the telson spine of later development.

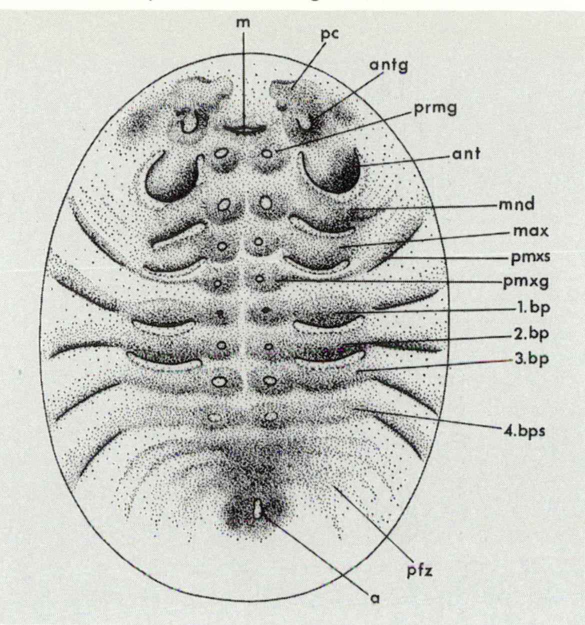

Fig. 8.1 *Ommatoiulus sabulosus* germ band with appendage buds, bp, leg pair; bps, leg bearing segment or somite; a, anus; ant, antennae; antg, antennal ganglia; m, mouth; max, maxillae (1st maxillae); mnd, mandibles; pc, protocerebrum; pfz, proliferation zone; pmxs, post maxillary segment or somite; pmxg, post maxillary ganglia; prmg, premandibular ganglia. Reproduced from Dohle (1974*b*) by kind permission of the author and the Zoological Society of London.

The pupoid enlarges quite rapidly once the shell has split. Newport (1841) likened this stage to the growth of a seed. Traces of antennae and leg primordia are visible.

In the next stage, the pupoid moults into the first true stadium which, in most cases, has three pairs of legs (hexapodous). Colobognath millipedes such as *Polyzonium germanicum*) have four pairs at this stage (Enghoff 1984*a*). These young millipedes are blind (even in species which eventually develop eyes) although the ocelli of the next stadium may be visible through the cuticle. Stadium I has fewer antennal segments than the adult and there is one pair of legs on each of segments 2, 3, and 4 (Causey 1943).

After a fairly short period of time (for example, 18–24 hours in *Orthomorpha* (=*Oxidus*) *gracilis*), stadium I moults to stadium II inside the egg capsule. This stadium has a simple ocellus (in species that are not blind as adults), more legs (usually six pairs), and one pair of defence glands.

The identity of the first active stadium to leave the nest depends on the species concerned. After a couple of hours, stadium I of *Polydesmus inconstans* (Polydesmida) is able to walk (Snider 1981*a*) and at this stage they break out of the nest by chewing a hole. In another polydesmid

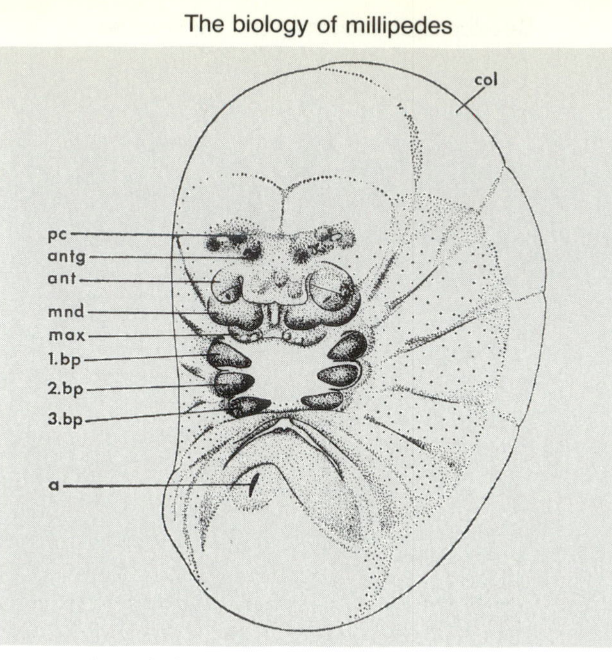

Fig. 8.2 *Ommatoiulus sabulosus* pupoid separated from the embryonic membrane. col, collum (for key to other labels, see Fig. 8.1). Reproduced from Dohle (1974*b*) by kind permission of the author and the Zoological Society of London.

species from Japan, Murakami (1965) noted that the hole is made in the side of the nest and appears to be engineered by a single animal. According to Demange and Gasc (1972), it is stadium III of *Pachybolus ligulatus* (Spirobolida) which leaves the nest and it is the same stadium which is the first active stage in *Narceus annularis* (Shaw 1966).

Whichever stadium it is that leaves the nest, it seems to be a general principle that the young millipedes only start feeding after moulting to stadium III. Thus, stadia I and II have to rely entirely on yolk for early development. The first meal usually consists of the empty egg capsule which may contain useful nutrients (Heath *et al*. 1974). Shaw (1966) speculated that bacteria left on the surface of the egg capsule by the female may inoculate the gut of the young millipede, serving to pass the gut flora between generations.

In *Glomeris marginata* in north west England, embryonic development takes about six weeks and the second stadium emerges from the egg capsule about eight to ten weeks after being laid (Heath *et al*. 1974). However, in the laboratory these development times can be altered by changing the temperature. For example, Juberthie-Jupeau (1974) submitted fertile eggs of *Glomeris marginata* to constant temperatures of between 4 and 28°C and found that development times ranged from 17 to 194 days with a minimum at 25°C (Fig. 8.3). Maximum percentage hatch was

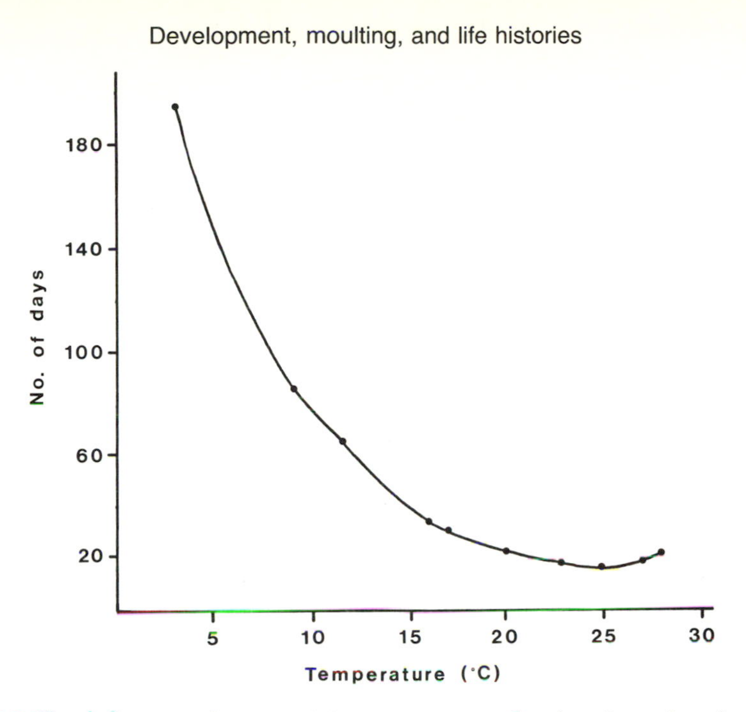

Fig. 8.3 The influence of constant temperature on the duration of embryonic development in *Glomeris marginata*. Redrawn from Juberthie-Jupeau (1974) by kind permission of the author and the Zoological Society of London.

between 16 and 17°C. When she subjected developing eggs to a thermal shock of 30°C (which proved fatal if maintained for a long period), malformations occurred in the segmentation of the germ band. These led to one, two, or rarely three extra pairs of legs at eclosion, with or without loss of a pair of limb buds and the appearance of an extra tergite (Fig. 8.4).

Detailed histological examinations were not performed on these malformed embryos. Thus, it was not possible to know whether the changes were due to (1) the appearance of an extra segment or diplosegment by abnormal segmentation of the germ band, or (2) disturbance of the sequence of embryonic development leading to the precocious formation of appendages and segments which are normally formed during post-embryonic development. Similar deformities following thermal shocks (i.e. one head and two abdomens) were noted by Aouti (1980). None of these malformed individuals were able to moult into stadium II.

The first three stadia of *Proteroiulus fuscus* are shown in Fig. 8.5.

8.3 Diplosegments

There is some controversy as to which diplopod segments are diplosegments, and from which segments certain structures are derived. The important

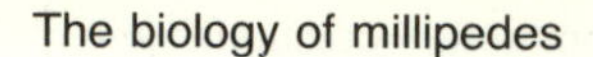

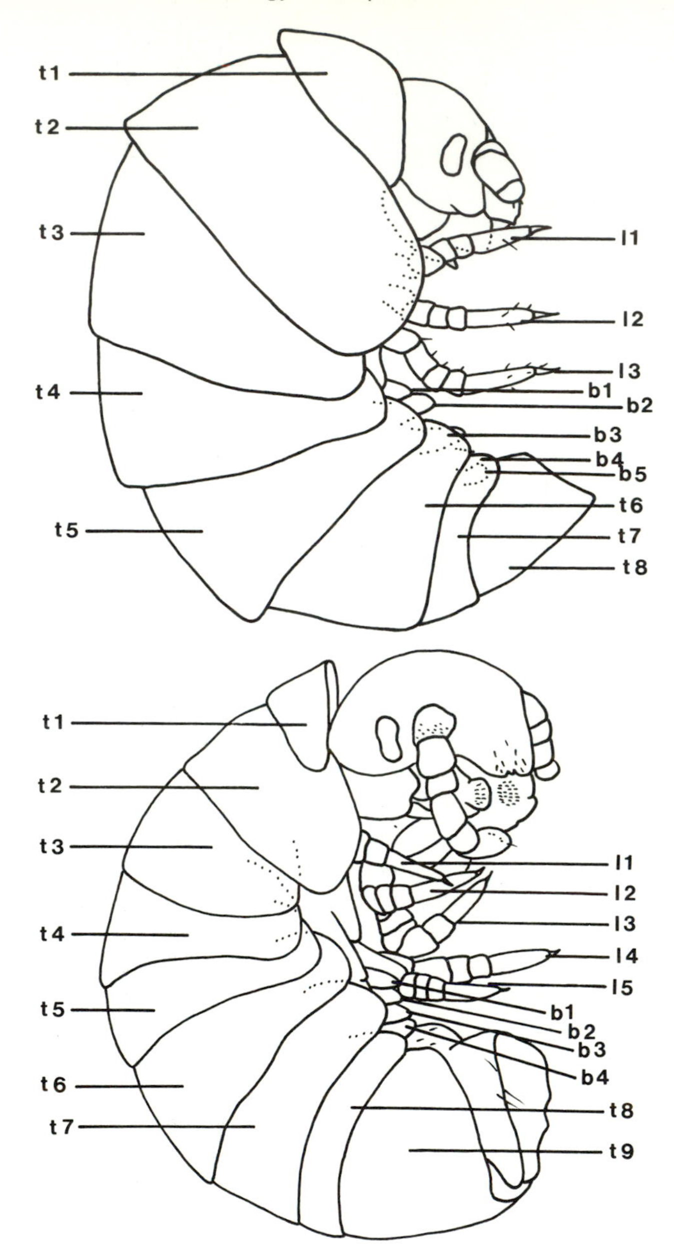

Fig. 8.4 Stadium I of *Glomeris marginata*. (a) Normal specimen with three pairs of legs (l1 to l3), five pairs of leg buds (b1 to b5), and eight tergites (t1 to t8). (b) Specimen subjected at the egg stage to a thermal shock. A malformed animal has developed with five pairs of legs, four pairs of leg buds, and nine tergites. Redrawn from Juberthie-Jupeau (1974) by kind permission of the author and the Zoological Society of London.

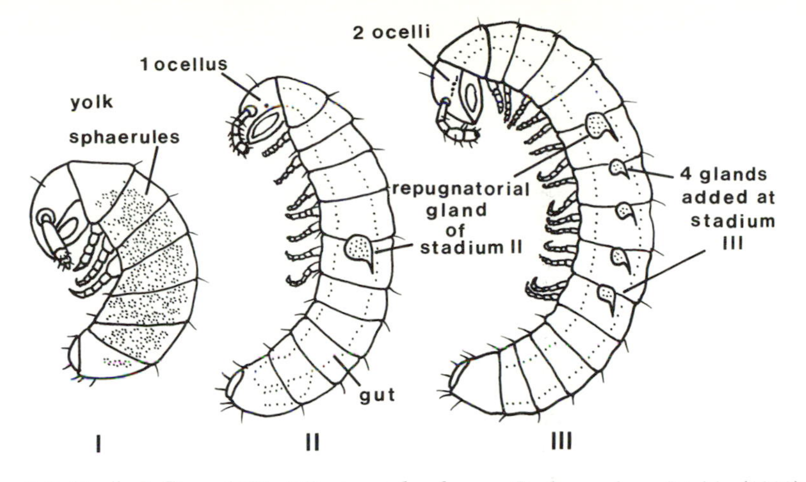

Fig. 8.5 Stadia I, II, and III of *Proteroiulus fuscus*. Redrawn from Dohle (1988) by kind permission of the author and the British Myriapod Group.

point to bear in mind is that segmentation observed in the germ band may not correspond directly with segmentation observed in the adult. In other words, there is an embryonic segmentation and a secondary segmentation which is imposed on it. Dohle (1974*b*) has provided evidence that the first leg-bearing segment forms the posterior part of the collum. The second leg-bearing segment forms the posterior part of tergite II, the third leg-bearing segment forms tergite III, the fourth forms tergite IV, and the fifth and the sixth leg-bearing segments unite in their lateral ectodermal parts to form tergite V, in other words the first double tergite.

Dohle's conclusions on 'segmentation' were based on embryological information whereas those of Kraus (1990) were made after studying exuviae (Figs 3.3, 3.4). Rings of moulted exoskeletons which were complete before ecdysis 'decompose' into discrete plates. True diplosomites have two tergites dorsally and two pairs of lateral pleurites. This separation occurs in the so-called 'thoracic segments' II to IV. Kraus interpreted this phenomenon as evidence that the anterior rings are diplosomites. The legs behind the head have been reduced in number to facilitate (1) rolling into a spiral, and (2) the flexing of the head to allow the charging of the gonopods of males with sperm before mating. This reduction (to avoid the dilemma of too many pairs of legs), has also resulted in a concomitant reduction in related internal structures, especially tracheal pockets and ganglia.

These questions are far from being resolved. Detailed studies are needed that involve serial sectioning of embryos to trace the movements of particular groups of cells during development, origins and tracks of nerves, muscle architecture, etc. We should finish this section with two quotes. 'The non-recapituation of a structure does not necessarily indicate its absence in phylogeny' (Kraus 1990). 'Adult morphology is not a good basis for drawing conclusions on the primary embryonic segmentation' (Dohle 1974*b*).

8.4 Moulting

Millipedes moult many times throughout their lives. Blower (1974*b*) found that more than 10 per cent of the life of *Ophyiulus pilosus* from egg to adult was spent in the process. The time of moulting is often synchronized with climatic conditions, particularly in older animals. *Narceus americanus* in North America times its annual moult to coincide with the dry period. The animals burrow into logs, form a chamber, and seal the entrance to conserve moisture. Prolonged desiccation has been shown to initiate moulting (O'Neill 1969). Polydesmids in Nigeria also moult in chambers in the dry season (Lewis 1971*a,b*).

The timing of moulting is undoubtedly under hormonal control but the identity of such hormones, and the site of their production, has not been determined. Nair (1980) reported the results of an experiment in which he implanted two pairs of cerebral glands of the millipede *Jonespeltis splendidus* (=*Anoplodesmus saussurei*) into fourth and fifth instar nymphs of the heteropteran insect *Dysdercus cingulatus*. This delayed the moulting of the insects, suggesting that a moult-inhibiting factor is released from the cerebral glands of millipedes which is common to insects. However, much more research is required on millipede neurosecretion before firm conclusions can be drawn about the mechanisms involved in moulting control.

Millipedes, like other arthropods, are at their most vulnerable when they are moulting. Most species seek a refuge in which to shed their exoskeleton. Some just make shallow depressions in the soil to moult in, or use existing structures, for example seed capsules (Snider 1981*a*), but others build elaborate chambers (Fig. 8.6). The walls of the chambers may be lined with faecal material (Causey 1943), or made from decaying wood or earth moistened with saliva. In the polydesmid *Euryurus erythropygus*, this material is moulded with the jaws and first pair of legs (Miley 1927).

Schlüter (1982, 1983) made an extensive study of the structure and function of the anal glands of *Rhapidostreptus virgator* (Spirostreptida). These produce a secretion which is used to bind faecal material together during construction of the moulting chamber. The gland complexes are found in the anal valves of both sexes. Each gland complex consists of about 200 secretory units, each of which comprises four cells: two secretory cells, an intermediary cell, and a canal cell. The amount of secretion produced by these glands varies during the moult cycle. It is very small in freshly moulted individuals, at a medial level during intermoult, and very large in the premoult phase.

Some moulting chambers are somewhat similar to the nests made for egg laying and may have a chimney-like projection. Toye (1967) described the

Fig. 8.6 Recently moulted *Orthoporus ornatus* excavated in its overwintering soil hibernaculum. Note the complete exuvium, part of which is usually eaten. Reproduced from Crawford *et al.* (1987) by kind permission of the authors, the Linnean Society of London and Academic Press.

formation of one such structure by a species of the polydesmid *Oxydesmus* (=*Coromus*) in Nigeria. The animals used semi-fluid excrement which hardened in contact with the air. The excrement was ejected from the anus and moulded into discs using the everted rectum and the anal valves. The discs were laid in overlapping rows. The last three pairs of legs held the rim of the nest while a fresh disc was positioned. The chamber was built up from the inside (rather like an igloo) with occasional strengthening material applied outside. The millipedes left the chamber periodically during construction to ingest fresh soil particles. The chambers were usually situated on firm foundations rather than directly on the soil surface and took from four to eight days to build. Building was faster at night.

The moulting process consists of four main stages (Halkka 1958). The first stage is the preparatory one described above where the millipede burrows into the soil or constructs a chamber. The second stage is the rigidation period, when the millipede lies on its side in a semi-circle. The body becomes about 15 per cent longer and the new integument can be seen between the old sternites which are pushed apart (see Figs 3.3, 3.4). The mouthparts protrude as do the gonopods and vulvae or rudiments thereof (Causey 1943).

During the third stage of moulting, ecdysis takes place. Internal pressure splits the old exoskeleton between the head and collum and the head and antennae are withdrawn from the old skin. A longitudinal split occurs also along the mid ventral line and, according to Miley (1927), also laterally above the legs. The body swells, the exuvia separates from the new cuticle and the animal 'walks out' of the old exoskeleton (Blower 1985; Causey 1943). The cuticular linings of the foregut and hindgut, and the epithelial cells of the midgut are also left behind (Hubert 1979*a*). In immature millipedes, extra segments are present in the animal that emerges from the old exoskeleton due to addition at the proliferation zone just anterior to the telson.

The fourth and final stage is that of recovery. The new cuticle is soft and pale and the legs and antennae are directed posteriorly. Slowly the appendages are bent, the cuticle darkens and hardens, and after three to four hours, the millipede starts to crawl. The exuvia is usually eaten and after a few days, the millipede emerges from its moulting retreat. The whole moulting process takes several weeks to complete.

Moults in younger millipedes tend to be of shorter duration than those in older ones. Snider (1981*a*) found that the final moult into the adult stadium took the longest time in *Polydesmus inconstans* (Polydesmida). She also recorded that moults took longer at lower temperatures.

Calcium salts are reabsorbed during premoult in most species. These are stored in the body (possibly the fat body and liver) and are remobilized to calcify the new exoskeleton.

8.5 Later development and life histories

8.5.1 Introduction

Millipedes develop by **anamorphosis**. As described earlier in this chapter, they hatch from the egg as a legless pupoid larva and then moult into a stadium I animal with few segments and usually three pairs of legs. Further moults follow, of varying number, when more segments are added (Fig. 8.5). Along with the segments are added legs, defence glands, and ocelli (if the species possesses eyes as adults).

At the posterior end of millipedes, just anterior to the telson, there are usually one or more segments which are apodous (without legs). These become the podous (leg-bearing) segments of the following stadium. In some groups, a maximum number of segments is reached after which there may be no more moults, or moults at which the length increases but not the number of segments (the latter being termed **hemianamorphosis**). In other groups, new legs and segments are always added at moult, even after sexual maturity is reached. This has led to the evolution of some complex life histories in certain species, particularly some of the Julida. This type of development contrasts with that of the insects which hatch with a full complement of segments (**epimorphic** development).

One point worthy of note here is the various methods of counting and expressing the number of segments. In julids, the most logical and perhaps most widely used method is to count the collum as segment 1, omit the telson, and express as podous segments plus apodous segments (e.g. head

+ 28 + 3 + telson, usually expressed as 28 + 3). Other workers might express the same millipede as 31–3 or 32/3. In contrast, for polydesmids, an absolute number is given which is the total number of segments including the collum and telson.

As invertebrates go, millipedes live for quite a long time. Whilst some are annuals, many take two, three, or four or more years to reach maturity. Some individuals of *Glomeris marginata* have been known to survive for 11 years (Carrel 1990). The reason for the long time taken for millipedes to reach maturity has been attributed to the poor quality of their food (Blower 1969).

8.5.2 Determination of stadia

8.5.2.1 Introduction

The stadia of millipedes are analogous to the instars of insects. As certain features change with each moult, it should be possible to determine which stadium an individual of a particular species is in. These include numbers of segments, defence glands, and ocelli, in addition to, of course, overall size and weight. In practice, however, stadial determination is not always as straightforward as it might seem.

Very many studies have been performed on the post-embryonic development of millipedes. There is insufficient space to describe them all but two have been selected to illustrate the main principles of stadial determination. The first section (8.5.2.2) concentrates exclusively on *Glomeris balcanica* as a representative of the pill millipedes. The second section (8.5.2.3) concentrates on *Julus scandinavius* but also includes a few details of stadial determination in other orders.

8.5.2.2 Determination of stadia in Glomeris balcanica

The pentazonids, such as species of *Glomeris*, are hemianamorphotic. They have two phases of development, an anamorphic stage, followed by an epimorphic one where moulting is accompanied by increases in weight and length, but not additional segments. *Glomeris balcanica*, a pill millipede from Greece, has been studied extensively in the laboratory and in the field by Iatrou and Stamou (1988, 1990*a,b*), on whose studies the following account is based.

Stadia in *Glomeris balcanica* can be determined up to stadium V by counting the number of segments (see Table 8.1). Several more moults take place before maturity is reached but there are no accompanying changes in morphological characters to allow easy stadial determination. Iatrou and Stamou (1988) finally managed to characterize these stadia by measuring the width of the second tergite. They were able to show that up to 17 stadia occur in the males and up to 19 in the females. Maturation occurred between stadia X and XIII. There are three separate periods of growth in *Glomeris balcanica*, the immature (anamorphic), the pseudomature (epimorphic but not mature), and the adult (sexually mature). In the latter two stages, the males and females can be separated.

8.5.2.3 Determination of stadia in Julus scandinavius *and other millipedes*

Julus scandinavius (Julida) is a common millipede in Britain. It was studied in detail by Blower (1970*a*) and Blower and Gabbut (1964). Most of the work was based on animals extracted using Tüllgren funnels, supplemented by pitfall captures, direct observation, and some laboratory studies. Identification of the stadia was attempted initially by simply measuring the length of each animal. These lengths were plotted on arithmetic probability paper which allowed stadia to be separated (Blower and Gabbut 1964). The method was improved by Blower (1970*a*) who extended the observations to include the number and arrangement of ocelli in the ocular field on each side of the head. The method he used was as follows:

In julids, stadium I millipedes lack ocelli, one ocellus appears in stadium II, two more in stadium III, three more in stadium IV, and so on. Subsequent rows are added at each moult (chordeumatids are similar, except that the second row consists of only one ocellus—Fig. 8.7). Thus, the stadium is equal to the number of rows of ocelli plus one. This method appears to work for many species that moult anamorphically (e.g. spirobolids—Spaull 1976) and is easy to use as long as the rows are not too reduced or jumbled.

Using such methods, Blower (1970*a*) was able to characterize the stadia of *Julus scandinavius* in a Cheshire woodland (Table 8.2). Other authors have produced similar tables for a wide range of other species since Blower's work. Some particularly detailed work by Peitsalmi (1981) gave all the combinations of segment numbers added at each moult for *Proteroiulus fuscus* in Finland.

Once one feels confident of being able to accurately determine the stadia of a particular species, other aspects of development can be studied. For example, Blower (1970*a*) was able to show that *Julus scandinavius* matured at stadium IX, X, or XI in its third year of growth. The sexes could be separated from stadium VII onwards when gonopods begin to develop on the seventh segment of the male. Details of the life histories of this and other species are given in Section 8.5.3.

The julids are completely anamorphic in their development—new segments are added at every moult. The polydesmids and chordeumatids develop in a similar fashion but have a fixed maximum number of segments. The polydesmids are blind and methods other than arrangement of ocelli have to be used for determining the stadia. The method usually used is simply to count the number of rings, since the number of segments added at each moult is quite regular, with little variation between individuals of

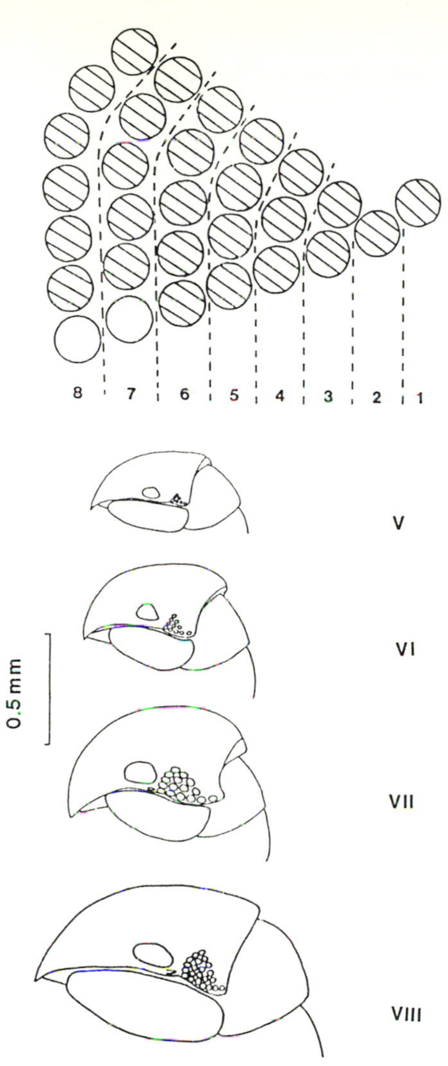

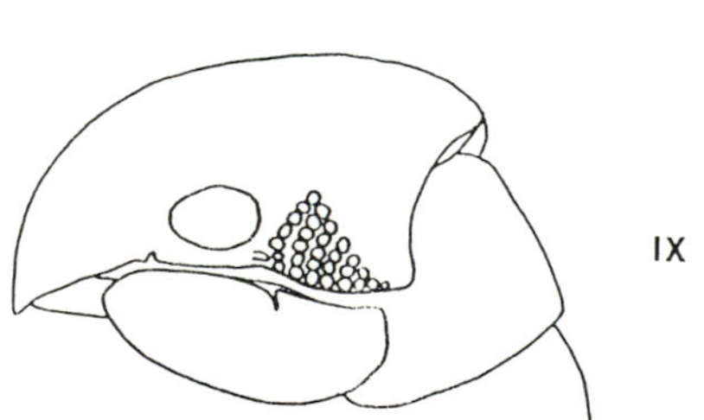

Fig. 8.7 (a) Diagrammatic representation of the ocular field of *Chordeuma proximum*. Shaded ocelli are those found most frequently in a British woodland. Those left unshaded are present in some animals. Numbers indicate the order in which the rows are added. (b) The ocular fields as seen in the animals from the left side (antenna removed) from stadia V to IX. Original drawing by H. Read.

Table 8.1 Morphological features of the first five stadia of *Glomeris balcanica*. After Iatrou and Stamou (1988). Reproduced by kind permission of the authors.

Stadium		No. apparent tergites	No. leg pairs	No. ocelli	No. antennal articles	Body length (mm ±SE)
I		8	8	3	4	2.85 ± 0.02
II		9	10	4	5	3.32 ± 0.02
III		10	13	5	7	4.10 ± 0.02
IV		11	15	6	7	5.40 ± 0.10
V	t	12	19	7	7	6.70 ± 0.11
	p	12	17	7	7	6.43 ± 0.19

Table 8.2 Morphological features of the stadia of *Julus scandinavius*. Segment numbers are those possessed by more than 10% of the population. After Blower (1970*a*). Reproduced by kind permission of the author.

Stadium		Podous segments	Apodous segments	Length (mm) x̄ ± 95% confidence limits	Volume (mm^{-3})
I		4	2		
II		6	5		
III		11	4		
IV		15	6	3.0 ± 1.6	0.85
V		21	6–5	6.4 ± 1.8	2.61
VI		26–28	6–5	7.8 ± 0.6	4.53
VII	Immature t	32–34	5–4	10.0 ± 2.3	8.99
	Immature p	32–34	6–5	10.6 ± 1.9	10.44
VIII	Immature t	36–38	5–3	13.3 ± 1.7	16.32
	Immature p	36–39	5–3	13.9 ± 2.2	19.31
IX	t	40–43	3–2	16.2 ± 3.1	28.63
	p	41–44	3–2	18.1 ± 3.0	37.77
X	t	44–46	2–1	20.8 ± 2.0	46.11
	p	45–47	2–1	23.8 ± 4.2	85.60
XI	t	45–47	2–1	22.5 ± 1.8	57.26
	p	46–47	2–1	26.5 ± 4.0	95.95

the same species (e.g. Banerjee 1973). Stadia of blind julids, or those with reduced numbers of ocelli, can sometimes be determined by using the pattern and number of defence glands (Halkka 1958).

8.5.3 Life history patterns

8.5.3.1 Introduction

Numerous papers have been published on the life histories of millipedes. From the early days of Newport (1841), this topic has held a fascination for myriapodologists. The direct observations of Peitsalmi (1981), Saudray (1953), Verhoeff (1923, 1926–32), and numerous others, have been complemented by the regular sampling programmes of Blower (1970*a,b*), Blower and Gabbut (1964), David (1982, 1984), David and Couret (1984), and Voigtländer (1987).

By far the largest number of studies on the life history patterns of millipedes have been carried out on members of the Order Julida. This is principally because species of julids are the commonest millipedes in temperate regions where most myriapodologists are concentrated! Life history patterns of julids are very diverse (see Read 1988). Notable studies have been those of Baker (1978*a,b*), Blower and Fairhurst (1968), Blower and Gabbut (1964), Blower and Miller (1974, 1977), Brooks (1974), David (1982), Halkka (1958), Peitsalmi (1981), and Sahli (1968).

Spirobolids and spirostreptids have been studied less often. Contributions have been made by Bercovitz and Warburg (1985), Demange (1972), Demange and Gasc (1972), Spaull (1976), Toye (1967), and Vachon (1947).

Polydesmids have been examined in temperate (Snider 1984*b*; Stephenson 1961) and tropical regions (Bhakat 1987, 1989*a,b*; Lewis 1971*a,b*). The most detailed studies on chordeumatids have been carried out in alpine regions (Meyer 1979; Pedroli-Christen 1978). See also Causey (1943) and Blower (1978).

The Penicillata (bristly millipedes) have been the subject of major works by Schömann (1956) and Meidell (1970), and more recently, Karamaouna (1990).

The Colobognatha are represented rather poorly with the only major contributions having been those of David and Couret (1983, 1984) and Murakami (1962, 1963).

In the Pentazonia, comprehensive work has been conducted on the temperate Glomerida (Iatrou and Stamou 1988, 1990*a*; Heath *et al.* 1974). However, the Sphaerotheriida and Glomeridesmida have hardly been examined at all.

So many studies have been made in this area of millipede biology that it would take a separate book to do them all justice. Before discussing some of the more unusual features of millipede life histories, the two rather different species (as discussed in Sections 8.5.2.2 and 8.5.2.3) will again be taken as examples.

8.5.3.2 Life history of Glomeris balcanica

In Iatrou and Stamou's (1990*a*) study on *Glomeris balcanica*, eggs were laid from May to August with a peak at the end of June. Hatching took place after two months. In this time, the eggs developed into the first free anamorphic stadium, I, via the pupal stage. Development through the anamorphic stages took about nine months and through the epimorphic stages a further 18 months. Thus Iatrou and Stamou (1990*a*) suggested a maturation time of two years six months in the laboratory; three years is probably more realistic in the field (Iatrou and Stamou 1991). This species goes on moulting and reproducing many more times before dying.

The work of Iatrou and Stamou indicated another feature that is common to many species of millipedes—the males often achieve maturity before the females (Iatrou and Stamou 1988). It is possible that more energy is used in the formation of the ovaries in females which retards maturation.

8.5.3.3 Life history of Julus scandinavius

Unlike Iatrou and Stamou's (1990*a*) study on *Glomeris balcanica*, which was primarily laboratory-based, that of Blower (1970*a*) on *Julus scandinavius* was based mainly on fieldwork. Blower was able to draw up a stadial spectrum for *Julus scandinavius* by counting the number of millipedes in each stadium in samples taken at monthly intervals throughout the year. In this way, a picture was built up of the life history of the species without having to follow individual specimens. As millipedes may live for several years, this was an important advance, although there are of course occasions when there is no alternative to direct observation.

The stadial spectra of *Julus scandinavius* in a Cheshire wood are shown in Fig. 8.8. From this it is clear that the new generation of millipedes appeared in the samples in June as stadium III. Laboratory stadies indicated that egg laying takes place mostly in April and about one month elapses between oviposition and emergence of stadium III animals. *Julus scandinavius* survives up to stadium III on yolk from the egg.

Throughout the summer, the juveniles grow and pass through four stadia to reach VI and VII by the winter (the latter being the first stadium in which the sexes can be distinguished).

After the first year, the development of *Julus scandinavius* is rather more difficult to follow. They start the second year of growth mostly as stadia VI and VII and probably reach stadia VIII and IX by the second winter. The millipedes then overwinter for a third time as stadia X and XI in the females or stadia IX, X, or XI in the males. The females mate and oviposit in their third spring and die.

The mortality rates in Blower's (1970*a*) study seem to have been quite high, certainly initially. Only 20 per cent of *Julus scandinavius* were found to survive the first month up to stadium III and only 1.5 per cent survived until the third winter. These estimates correspond well to those of Levieux and Aouti (1978) who worked on a rather different species *Pachybolus laminatus* in the Ivory Coast. A life table was produced which revealed a high overall larval mortality of 97–99 per cent. Most mortality occurred in stadium VI.

8.5.4 Periodomorphosis

Periodomorphosis is the extension of life of the male by means of intercalary or Schalt stadia. In some species of millipedes, the male is able to moult from a

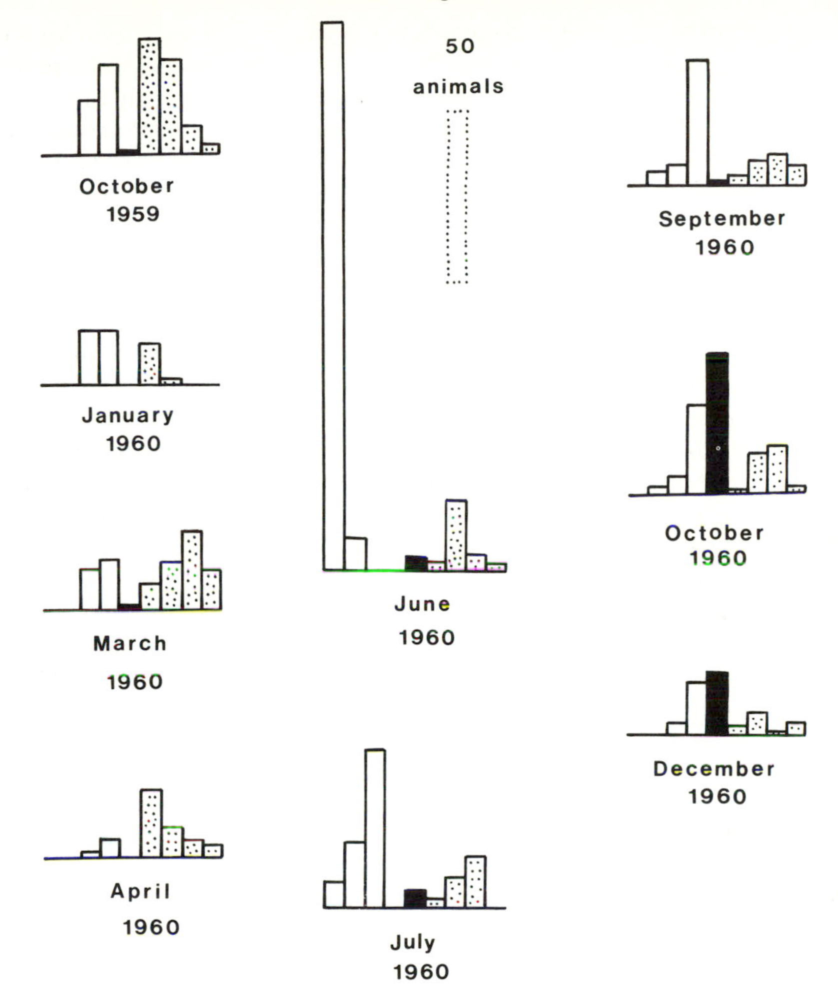

Fig. 8.8 Numbers of *Julus scandinavius* found in Tüllgren samples from a deciduous wood in Britain. Open columns represent stadia III to VI, the black column stadium VII and the stippled columns stadia VIII to XI. Redrawn from Blower (1970*a*) by kind permission of the author and the Zoological Society of London.

fully-developed sexually active stage, into an intercalary stadium where the secondary sexual characteristics (e.g. gonopods and modified first pair of legs) are intermediate between immature and mature, and are not fully functional. This was referred to as the 'eunuch' stage by Baker (1979*a*). A further moult is required to achieve full mature status again.

Intercalaries were first described by Verhoeff (1923) who thought that they occurred between the juvenile and mature stages. Subsequently,

Verhoeff (1926–32) found them to occur between male moults. This has since been confirmed (Sahli 1958*b*; Blower and Fairhurst 1968). Verhoeff (1926–32) proposed that in the julid *Tachypodoiulus niger*, maturity was reached in stadium VIII, followed by alternation of intercalary and mature stages (summarized in Blower and Fairhurst 1968). However, the sequence is not always as regular as this; an intercalary can moult into another intercalary (Halkka 1958; Sahli 1958*b*, 1961*b*).

Periodomorphosis seems to occur predominantly (but not exclusively) in Julida of the genera *Ommatoiulus* and *Tachypodoiulus*. Intercalary males were described in *Brachydesmus superus* (Polydesmida) by Stephenson (1961), but these were overwintering stages before the attainment of full maturity and were not true intercalaries.

The function of true intercalary stadia has been much debated. Another julid, *Allajulus (Cylindroiulus) nitidus*, has intercalary stadia in some populations but not in others. The males are able to moult directly from one mature stadium to another. Sahli (1985*b*) suggested that intercalary stadia in this species occur when environmental conditions are poor. Fairhurst (1974) proposed that intercalaries were resting stages for the build up of sperm (citing Mauriès unpublished), but this must also be as necessary in *Allajulus nitidus*. Blower and Miller (1977) suggested that an inactive stage in, for example, *Tachypodoiulus niger*, suppressing the hormones causing copulatory stages to wander, would minimize the risk of predation. This would not be so important in *A. nitidus* as it is subterranean.

Periodomorphosis has also been proposed as a method of correcting imbalances in the sex ratios. Blower and Fairhurst (1968) and Fairhurst (1974) stated that in a vagile, actively dispersing species, lengthening the male life secures a higher chance of breeding in new areas. Baker (1978*b*) proposed that periodomorphosis, as well as 'spreading the risk' over areas by keeping an even sex ratio, also does so over time. Thus, after a good breeding year, there is little need for periodomorphosis as the next generation is secured. A poor breeding season may not ensure continued survival, so the animals can then extend their life and breed again. Sahli (1986) has also supported this idea of insurance against bad breeding conditions. However, the function of intercalaries is still far from being understood. More work is required on the hormones of millipedes to elucidate the signals that determine when a male moults into an intercalary stage.

Halkka (1958) stated that periodomorphosis (as manifested here) is confined to the Diplopoda. However, there are examples of similar phenomena in other invertebrates. These include several species of Crustacea such as decapods of the genus *Cambarus*. Two forms of male can be distinguished which differ in secondary sexual characteristics, particularly the organs of sperm transfer. Moulting from a sexually functional form to one lacking the ability to reproduce, and back again, has been shown to occur.

8.5.5 Parthenogenesis

Several species of millipede can reproduce parthenogenetically. Some species are known to have a very low ratio of males to females, or no males at all. This condition is termed **spanandry**. Spanandry may occur in many species at certain times of the year or in specific habitats, but there are identifiable occasions when it is likely to indicate parthenogenesis.

Thelytoky is the production of female offspring from virgin mothers and is the form of parthenogenesis found in millipedes. Enghoff (1978*a*) listed the instances of thelytoky and spanandry in the different orders of millipedes. He noted that among the julid millipedes the sole examples are from the family Blaniulidae, whereas in the polydesmids examples are much more widespread.

Three species of millipedes have been studied extensively with regard to parthenogenesis. These are the bristly millipede *Polyxenus lagurus*, and the two julids *Nemasoma varicorne* and *Proteroiulus fuscus*. *Proteroiulus fuscus* appears to be parthenogenetic in virtually all its range. Males do occur in some populations. Rantala (1974) found males were abundant in some laboratory populations (up to 20 per cent compared with only 1 per cent in field populations) but it seems likely that they are non-functional (Enghoff 1978*a*). In contrast, *Polyxenus lagurus* and *Nemasoma varicorne* show regional variation in the degree of spanandry.

Schömann (1956) considered there to be two separate races of *Polyxenus lagurus*, bisexual and parthenogenetic. These two forms differ morphologically and have been mapped to show their distribution across Europe (Meidell 1970; Enghoff 1976*a*).

A similar situation exists with *Nemasoma varicorne* and has been studied by Enghoff (1976*b,c* 1978*b*). The vulval receptaculum of the parthenogenetic female in this species is not fully developed, in contrast to those from bisexual populations. The latter are unable to reproduce successfully in the absence of males. Generally, the bisexual female produces more eggs but the males (in the sexual populations) are more susceptible to desiccation. The distribution of the two forms is shown in Fig. 8.9.

The distribution of both forms of *Nemasoma varicorne* corresponds with the deciduous forest zone. The bisexual form occurs mainly in the centre whereas the parthenogenetic form is more peripheral. The boundary between the forms is narrow and passes through Denmark where the distributions have been mapped in detail by Enghoff (1976*b*). The boundary relates to changes in soil type and other factors such as the distribution of ancient deciduous forest and barriers such as the Baltic Sea. Within the same forest, the two forms tended to be found on different trees but Enghoff (1978*b*) could not relate any obvious environmental factors to the distribution.

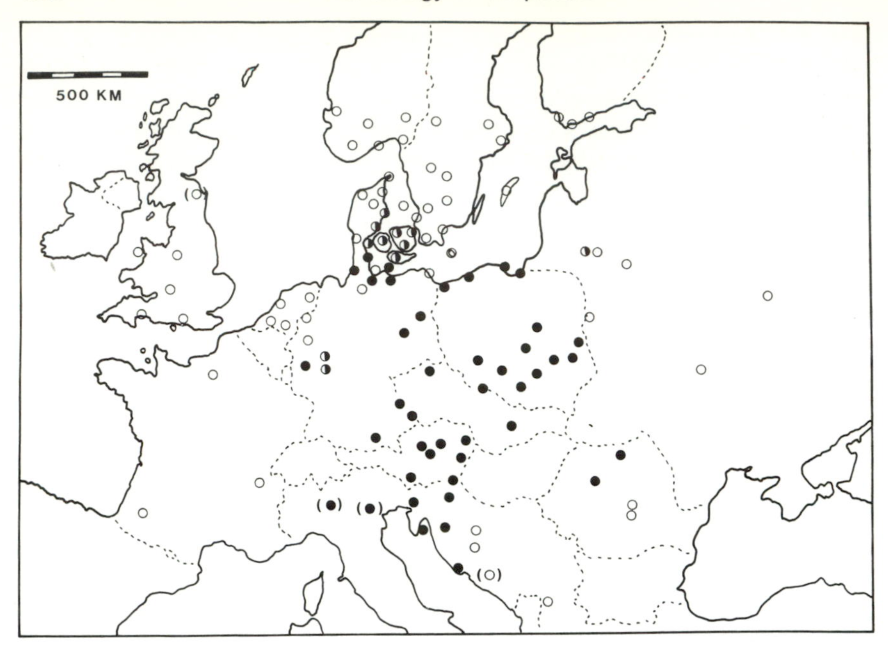

Fig. 8.9 Total distribution of *Nemasoma varicorne*. Bisexual form (●), thelytokous form (○) and both forms (◑). The exact localities of the symbols in brackets are unknown. The species also occurs in Scotland, Belgium, Luxembourg, Hungary, Bulgaria, and a little further east in the USSR than shown on the map. Redrawn from Enghoff (1978*b*) by kind permission of the author and the Danish Natural History Society.

8.5.6 Environmental factors: climate and altitude

There are many examples of close association between climatic factors and the activity times and breeding seasons of millipedes. In temperate regions, many species hibernate during the cold winter months. For example, *Blaniulus guttulatus* moves down into deeper soil layers to hibernate when the temperature falls below 5 °C (Biernaux and Baurant 1964; Brookes and Willoughby 1978). In desert areas, the hot dry season may be spent in aestivation. Development only occurs when there is moisture (Crawford *et al*. 1987). Because of the short growth period in extremely arid areas, millipedes such as the spirostreptid *Archispirostreptus syriacus* may take up to nine years to mature.

Individual species may be affected very specifically by climatic factors. Banerjee (1967*a,b*) described a complex migration pattern for *Cylindroiulus punctatus* (Julida) in Britain. When the weather starts to warm up in the spring, the species moves from the leaf litter to rotting logs. Mating and oviposition take place in the wood. When the temperature begins to drop again in the autumn, the millipedes move back into the leaf litter and eventually down to the mineral soil where they spend the winter.

Table 8.3 Life history characteristics of Choreumatida and associated climate parameters in the central Alps, Tyrol. After Meyer 1990. Reproduced by kind permission of the author.

Species	*Ochogona caroli*	*Ochogona caroli*	*Ochogona caroli*	*Trimerophorella nivicomes*
Altitude (m)	670	1470–1750	2000	2500
Time with temp. $> 5°C$ ($h\ y^{-1}$)	4300	?	1800	780
Snow free season	Mar–Dec	Apr–Nov	May–Oct	June–Oct
Breeding season	April	May	June	July/August
Mean time to maturity (months)	17	18	27	37
Life cycle (years)	2	2	3	4

Some interesting work by Meyer (1979, 1985, 1990) has tied together some of the factors that affect the life cycles of millipedes. The studies concentrated on chordeumatid millipedes of alpine regions. Meyer (1990) summarized the life histories of four species found at a range of altitudes (Table 8.3). With an increase in altitude, decrease in snow-free days, and decrease in temperature, there is a corresponding increase in time to maturity, longevity, and a general slowing down of the life cycle. The species that lives in the high Alpine zone, *Trimerophorella nivicornes*, is clearly adapted to a very special habitat where there are extreme daily fluctuations in temperature and a very short snow-free period. This species takes four years to complete its life cycle. Related species at lower altitudes take only two years and some populations outside the Alps are annual. Such extension of the life history from one to two years has also been recorded by Pedroli-Christen (1978) in different populations of *Craspedosoma alemannicum*.

Blower (1969) considered that the ability to vary the age or stadium of maturity offers a degree of flexibility in the species so that it can adapt to conditions prevailing at the time. David (1991) raised the same point but questioned whether it is stadium or age which is critical. Taking three julid millipedes as examples: females of *Allajulus (Cylindroiulus) nitidus* mature at four years but at stadium IX in Britain and stadium X in France; in contrast, *Ophyiulus pilosus* and *Julus scandinavius* both mature at stadium X but take two and three years respectively to reach it.

Altitude and climate were considered to have an effect on the life cycle of *Ommatoiulus moreleti* in Madeira (Read 1985). Millipedes at low altitudes initially grow more rapidly since it is warmer there than at higher altitudes. In the summer months, the lower regions become hot and dry whilst higher areas become warmer and wetter. Growth at higher altitudes then catches up with that of millipedes in coastal regions.

8.5.7 Fecundity

The number of eggs laid by a single female may vary considerably (both between and within species) and may be related to temperature (Snider 1981*b*). Nests produced by female *Streptogonopus phipsoni*, for example, may contain between 129 and 731 eggs (Bhakat *et al.* 1989). Other polydesmids may produce as many as 2000 eggs in a lifetime (Snider 1981*b*). These can be laid in several nests and may all be fertilized by sperm stored by the female after a single insemination.

Other species survive to mate more than once and produce two or more separate broods (iteroparous). *Glomeris marginata* for example may live for up to ten years, and produce more than seven broods in a lifetime (Blower 1985). Since an individual female can produce more than 80 eggs per year (Heath *et al.* 1974) it is clear that over its lifetime, a single *Glomeris marginata* can produce as many eggs as the apparently more fecund polydesmids, albeit spread over a longer period.

Considerable within-species variation may exist in the number of eggs in a nest. In *Orthoporus ornatus* Crawford *et al.* (1987) described how individual eggs of this desert species are coated with faecal material, then deposited in clutches of anything between 50 and 500. Size is also important. Large female *Glomeris marginata* lay more eggs than small females (Heath *et al.* 1974).

8.5.8 Variety of life histories

Various different types of life cycle can thus be identified in millipedes, some of which are shown in Table 8.4. These different categories, coupled with the varying lengths of time taken to reach maturity, indicate the wide range of life histories that occur. What are the reasons for this tremendous variety? Blower (1969, 1970*b*), and more recently David (1991), have attempted to unravel some of the complexity. However, more work is needed that relates physiology to ecology in the field before this question can be answered.

8.6 Life histories and evolution

Blower (1969, 1970*b*) attempted to relate the life histories of British millipedes to their habitats and habits. For example, semelparous species of Julida (i.e. those breeding only once) such as *Julus scandinavius* and *Ophyiulus pilosus* are found in woodland leaf litter. Consequently, they are distributed evenly throughout the forest floor. In contrast, iteroparous species of Julida (i.e. those able to breed more than once) such as *Cylindroiulus punctatus* require rotting logs for feeding and oviposition. This species has a more patchy distribution.

Calow (1978) has suggested that iteroparity may be an insurance policy against bad reproductive years. Thus animals with predictable habitats and resources are less likely to suffer reproductive disasters than those which de-

Table 8.4 Summary of life history features in millipedes. IC, intercalary; ♂–♂ mature males moult directly to another male without passing through an intercalary.

	Female reproduction			Male reproduction			Example
	Breed once	Breed more than once	Parthenogenesis	Breed once	IC	♂–♂	
Semelparous							
	✓			✓			*Ophyiulus pilosus*
Iteroparous							
Intraseasonal		✓		✓			*Polydesmus angustus*
Interseasonal		✓		✓			*Cylindroiulus punctatus*
		✓			✓		*Tachypodoiulus niger*
		✓				✓	*Allajulus nitidus*
		✓	✓		✓		*Proteroiulus fuscus*
		✓	✓			✓	*Nemasoma varicorne*

pend on scattered resources. This fits in with Blower's comments concerning the millipedes. The same result can be achieved by varying the development or maturation period so that the breeding season is spread over a long time.

David (1991) considered iteroparity and semelparity to be more related to taxonomic position than the habits of different species. He argued that 'such interspecific' differences are unlikely to originate from purely environmental factors associated with species' habits. Individuals of *Polydesmus angustus* (Polydesmida) from the same nest, reared in the same temperature regimes, may take either one or two years to mature. According to David, these variations of age and stadia at maturity, and also number of rings, are more likely to have a genetic basis. However, there must have been selection for the optimum life history for each species.

Another point made by Blower (1969) concerned a pair of common British species, *Julus scandinavius* and *Ophyiulus pilosus*. *Julus scandinavius* takes three years to reach maturity, is large in size, and produces many eggs. In contrast, *Ophyiulus pilosus* matures in two years, is smaller, and lays only half the number of eggs of *Julus scandinavius*. The result of this is that *Ophyiulus pilosus* probably has a higher intrinsic rate of increase and a more contracted life history, and is probably able to increase in numbers more rapidly. Blower (1969) referred to such species as 'neotenous' and pointed out that many millipedes in this category have been successful colonists. *Ophyiulus pilosus* has indeed spread to many countries.

Neoteny was also discussed by Demange (1974). He pointed out that neotenous species were characterized by the addition of only a small number of segments at each moult as well as taking fewer moults to achieve maturity (and therefore a reduced number of segments in the adults). In addition, they are often blind as adults. In this context, whole groups such as the Polydesmida and Chordeumatida are neotenic in comparison to the Julida.

Apart from neoteny, several other features of the life cycles of millipedes are possible prerequisites for colonization. Notable amongst them is parthenogenesis, which may aid the successful spread of many species from single females.

One of the most renowned colonizers is *Ommatoiulus moreleti*. This species is indigenous to the Iberian Peninsula but has spread successfully to South Africa, various islands, and to Australia where it has become a considerable pest (Baker 1979*a*—see also Section 10.5). *Ommatoiulus moreleti* undergoes periodomorphosis. Its longevity, coupled with its considerable wandering ability, has aided dispersal and colonization. Larger (presumably older) animals have been recorded frequently in extreme habitats, at the tops of mountains for example. This species exhibits a very different set of characteristics from Blower's (1969) neotenous species, which are more classical *r*-strategists. It also demonstrates the difficulties inherent in generalization. *Ommatoiulus moreleti* presumably has a very flexible life cycle and is able to adapt quickly to new circumstances.

9

Predators, parasites, and defence

9.1 Predators

There is surprisingly little quantitative information on the numbers of millipedes that fall victim to predation. Most information derives from anecdotes, or when millipedes turn up in the gut contents of vertebrates. A wide range of invertebrates, amphibians, reptiles, birds, and mammals will take millipedes in captivity (Baker 1985*b*; Remy 1950). However, it is by no means clear that the same predators will take them in the wild.

The early stadia must be much more vulnerable than the adults, but there is very little information on predation of juveniles. Baker (1974) reported that early stadia of *Brachydesmus superus* are eaten by the mite *Pergamasus quisquiliarum*. Numerous eggs must also fall victim to soil scavengers.

Considering the wide range of defensive chemicals that are present in millipedes (see Section 9.3) it is perhaps surprising that so many fall victim to predation. The sedative secretions of *Glomeris marginata* provide effective defence against most invertebrates and vertebrates, with the exception of hedgehogs which eat many pill millipedes in the wild (see Carrel 1990 for references). In addition to hedgehogs (Dimelow 1963), other vertebrate predators of millipedes include elephant shrews (Wood *et al.* 1975), water shrews (Churchfield 1979), frogs (Stachurski and Zimka 1968), lizards (Sadek 1981), and turtles (Wheeler *et al.* 1964). Some interesting co-evolutionary strategies are sure to exist between particular millipede species and predators and this topic would repay further study.

The African pill millipede *Sphaerotherium* rolls into a tight ball as an effective defence against ants, birds, and mice. However, it is no defence against the mongoose, which smashes the millipedes against rocks and then eats them (Fig. 9.1). Some doubt has been expressed as to the veracity of this behaviour. However, Eisner (1968) related a passage from a book by D. Wager (1946) *Umhlanga—a Story of the Coastal Bush of South Africa* (Knox, Durban, South Africa) in which the behaviour is described in graphic detail. The passage states:

> Mongooses in captivity eat almost anything, but in their wild state, they live mainly on insects. A friend of mine recently told me a strange tale about one of these creatures. He's an old man, and he's more or less grown up in these wild

Fig. 9.1 Three consecutive stages in the hurling and smashing of *Sphaerotherium* by the banded mongoose. Reproduced from Eisner and Davis (1967) by kind permission of the authors and the American Association for the Advancement of Science (AAAS). Copyright 1967 by the AAAS.

stretches of Natal. He said that one morning when he was sitting quietly under a tree in the bush hoping to see some birds, he spotted a colony of mongooses nearby. Suddenly, one of them climbed a short distance up a tree and knocked down a pill millipede. The mongoose jumped down after it, grabbed it between his front feet, and hurled it through his back legs against the tree. The impact smashed the otherwise impregnable ball, and before any of his friends could cheat him of his prey, he ate it.

Millipedes may form a major component of the diet of carnivorous beetles. Snider (1984*a*) reared a range of carabid and staphylinid beetles from egg to adult on *Polydesmus inconstans* and *Ophiulus pilosus* (both introduced species in Michigan). Female beetles were subsequently able to oviposit. The devil's coach horse beetle *Staphylinus olens* will eat *Ommatoiulus moreleti* (Baker 1985*b*). Beetles will also eat millipedes in deserts, as will scorpions (Crawford *et al.* 1987).

Although most ants appear to be deterred by the defensive secretions of millipedes, there are at least five species which will consume them (Hölldobler and Wilson 1990). Included among these is *Myopias julivora* from New Guinea, a specialist predator of penicillate (bristly) millipedes (Willey and Brown 1983).

9.2 Parasites

9.2.1 Introduction

Millipedes may be hosts to a wide range of invertebrate parasites, many of which are species-specific. Some studies have included an applied aspect in searching for biological control agents for millipedes which are pests. Others have examined the roles of millipedes as intermediate hosts for vertebrate parasites. All in all, this is one of the better-understood areas of millipede biology. In this section, the main parasites of millipedes will be described. Immune responses to blood infection are described elsewhere (Section 5.3).

9.2.2 Fungi

Some species of *Cylindroiulus* are attacked fairly frequently by ectoparasitic fungi of the Laboulbeniales (Rossi and Balazuc 1977). The hyphae infect the anterior-most three pairs of legs of females, and anterior-most seven pairs of legs of males. Blower (1985) suggested that this restricted distribution may be due to the millipedes being unable to groom these legs effectively, or to the absence of defensive glands from the first five rings.

9.2.3 Iridoviruses

These viruses get their name from the appearance of infected invertebrates. They 'glow' with a violet sheen. This is due to refraction of light by the tiny virus particles beneath the cuticle. In terrestrial isopods, infection with iridoviruses is quite common (Federici 1984) and some types will cross-infect between lepidopteran larvae, isopods, and centipedes (Ohba and Aizawa 1979). However in this study, the authors were unable to infect *Oxidus gracilis* or *Rhysodesmus semicirculatus*. Nevertheless, one of us (S. P. H.) has occasionally observed specimens of *Cylindroiulus punctatus* in the field with a distinct violet sheen, apparently infected with an iridovirus.

9.2.4 Rickettsia

Wilson and Burke (1972) reported that a couple of days after bringing specimens of *Ophyiulus pilosus* into the laboratory, all the millipedes died. Tissue smears of the fat body were light blue in appearance and contained a coccobacillus-shaped organism which was present frequently in long chains. The average size of these microorganisms was 400 × 265 nm, although some were up to 800 nm in length. Rickettsial-like microorganisms have also been detected in the Malpighian tubules of *Polyxenus lagurus* where they were disease organisms rather than symbionts (Schlüter and Seifert 1985*b*).

9.2.5 Protozoa

Infection by gregarines is quite common and widespread in millipedes (e.g. Crawford *et al*. 1987; Heath *et al*. 1974). Gibbs (1952) studied the incidence of a sporozoan blood parasite of the South African garden millipede *Archiulus* (=*Ommatoiulus*) *moreleti*. The incidence of infection varied from 1–50 per cent over a few months and it was sometimes difficult to find an uninfected cell in the blood. Severe infections resulted in the millipedes developing 'gangrene' in their legs which eventually decayed, leaving stumps. This parasite was more recently given the name *Gibbsia archiuli* by Levine (1986) who studied its life cycle in greater detail. Merogony, gamogony, and sporogony all take place in the blood cells of the millipede. The oocysts contain four sporocysts, each with one sporozoite, and the microgametes are not flagellated.

9.2.6 Acanthocephala

Bowen (1967) performed a series of experiments on the development of the acanthocephalan *Macracanthorhynchus ingens* in the spirobolid millipedes *Floridobolus penneri* and *Narceus gordanus*. The millipedes are intermediate hosts of the parasite, the adults of which live in racoons. The acanthors were seen in the posterior portion of the midgut in both species of millipede. The parasites penetrated the gut epithelial cells which were destroyed completely. A thin fibrous capsule was formed around each parasite by blood haemocytes as it moved into the haemocoel from the gut cells.

9.2.7 Nematophora

Infection by nematophorans has been reported occasionally (e.g. *Gordius* in *Ommatoiulus moreleti* by Baker 1985*a*). Sahli (1972) reported that such infection in *Tachypodoiulus niger, Ommatoiulus sabulosus*, and *Cylindroiulus teutonicus* resulted in loss of secondary sexual characters which he termed 'castration parasitaire'.

9.2.8 Nematodes

Blower (1985) considered nematodes to be the most common endoparasites of millipedes (Fig. 9.2). They have been found in a wide range of species all over the world (Adamson 1987*a,b*; Bowie 1985; Crawford *et al.* 1987; Dollfus 1952; Upton *et al.* 1983; Wright 1979). Special interest has centred on the nematode *Rhabditis necromena* as a possible biological control agent for the millipede *Ommatoiulus moreleti*, a pest in Australia. Interestingly, after an initial population explosion, this millipede appears to be declining (see Section 10.4). It has been suggested that nematode infection may be the cause (McKillup and Bailey 1990; McKillup *et al.* 1988).

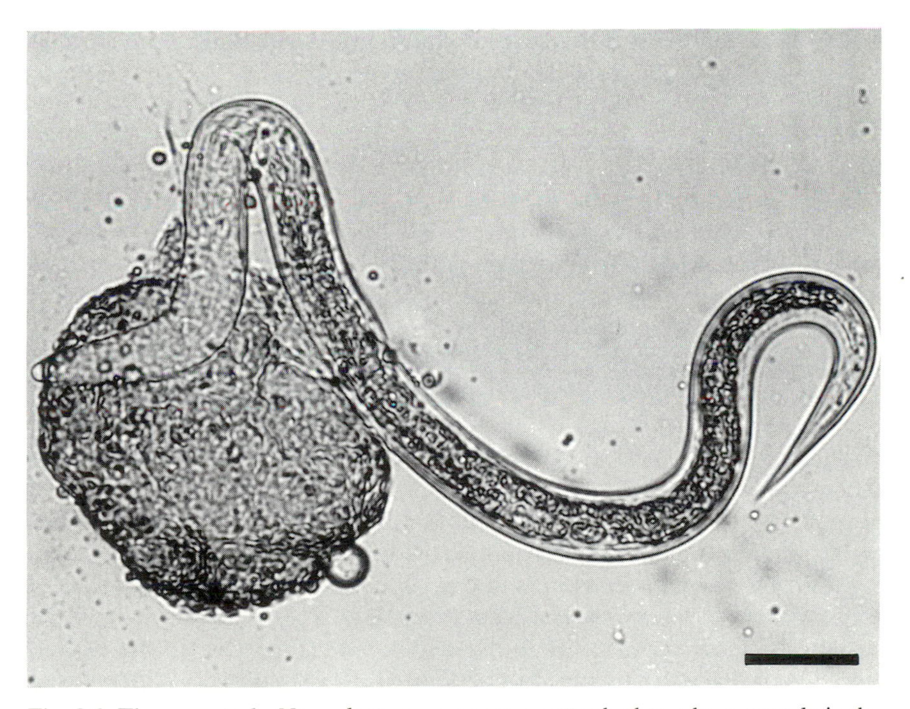

Fig. 9.2 The nematode *Neoaplectana carpocapsae* attached to a host capsule in the haemocoel of *Oxidus gracilis*. Scale bar = 50 μm. Reproduced from Poinar and Thomas (1985) by kind permission of the authors and Academic Press.

In Australia, the normal host for *Rhabditis necromena* is a native millipede but it has clearly been able to infect the introduced species also. Juveniles of the nematode remain inside the haemocoel of the millipede until it dies, then resume their development after feeding on bacteria present in the decaying carcass. Death of *Ommatoiulus moreleti* appears to be caused by soil bacteria which proliferate inside the millipede after entering via the surface of the nematodes (Schulte 1989*a*). *Ommatoiulus moreleti* does not seem to possess immunity to these bacteria in the same way that species of native millipede do. *Oxidus gracilis* can also be killed as a result of nematode infection, but again it is bacteria introduced at the time of infection which are the lethal agents rather than the nematodes themselves (Poinar 1986; Poinar and Thomas 1985).

In a phylogenetically ancient group such as millipedes, there has been considerable time for complex host–parasite relationships to develop. For example, Adamson and Van Waerebeke (1985) reviewed the geographical and host distributions of *Rhigonema* and concluded that this was probably a paraphyletic group which included primitive members of several different lineages. In a detailed study of the nematode fauna of *Rhinocricus bernhardinensis*, Adamson (1985) found seven species co-existing in the posterior intestine. Three superfamilies were represented. While the Rhigonematoidea were relatively *K*-selected, Adamson considered the Thelastomatoidea and Ransomnematoidea to be more *r*-selected.

9.2.9 Diptera

Enghoff (1976*a*) reported finding a single larva of a gall midge on *Polyxenus lagurus*. Apart from this, the only studies of which we are aware have again been on *Ommatoiulus moreleti* in the search for a biological control agent (McKillup and Bailey 1990). The millipedes' native area is Portugal and a search was made for naturally occurring parasites with a view to introducing these to Australia to control *Ommatoiulus moreleti*.

The parasite studied was eventually identified as a sciomyzid fly *Pelidnoptera nigripennis* (Bailey 1989). This has been transferred recently to the Phaeomyiidae (Vala *et al.* 1990). The adults lay eggs on the millipedes during the spring in southern Portugal and the first instar larvae penetrates through the soft tissues of the millipede (Fig. 9.3). The parasitoid spends the summer as a first-instar larva, kills the host during the autumn when in the third instar and overwinters in the pupal stage. The maximum rate of parasitism observed was about 35 per cent (Fig. 9.4). The life cycle of the parasite is closely tuned to that of the millipede (Fig. 9.5). In the laboratory, 15 species of millipedes from 5 families were exposed to the parasite but only Julida were successfully parasitized.

Approximately 1000 millipedes parasitized by *P. nigripennis* adults collected from around Lisbon, Portugal, have been imported to Australia.

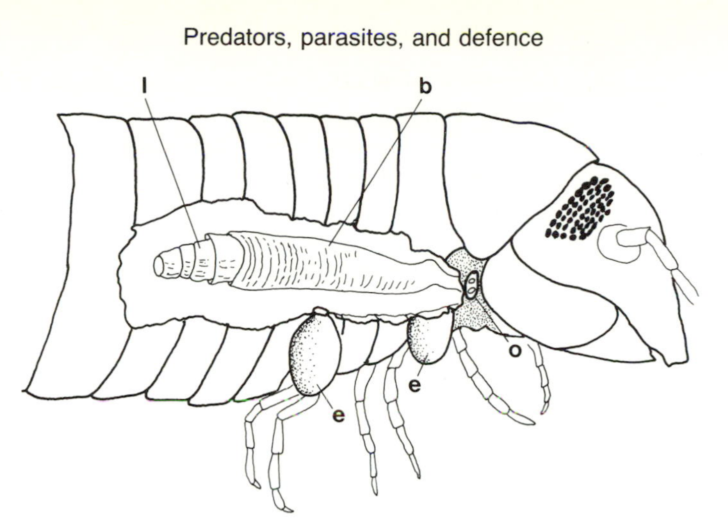

Fig. 9.3 *Ommatoiulus moreleti* showing empty eggs (e) of *Pelidnoptera nigripennis* and the first instar larva (l) with most of its body within the breathing tube (b) which opens (o) through the intersegmental membrane (stippled area). The head of the millipede is slightly extended. Redrawn from Bailey (1989) by kind permission of the author and CAB International Institute of Entomology.

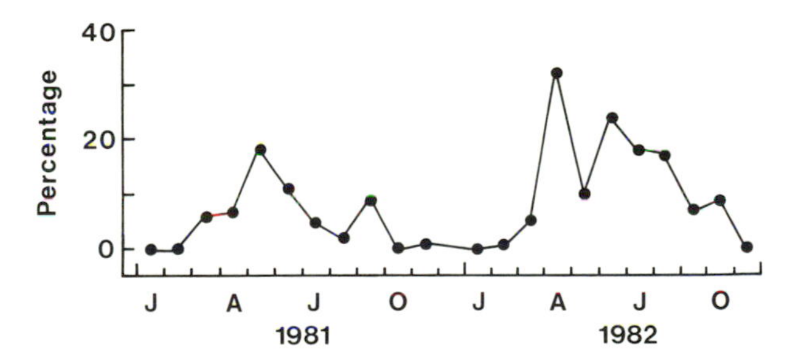

Fig. 9.4 Percentage of *Ommatoiulus moreleti* (stadium VIII and older) parasitized by *Pelidnoptera nigripennis* at Cascais each month during 1981–1982 (N for each month ≥ 29). No samples were taken during December 1981. Redrawn from Baker (1985*a*) by kind permission of the author and CSIRO Australia.

Following specificity testing, these have been released in wooded areas around Adelaide, South Australia, which has a similar climate to Lisbon (Vala *et al.* 1990). Results are awaited!

9.2.10 Mites

'Phoresy' is the phenomenon whereby small invertebrates with relatively low powers of migration 'hitch a ride' on a larger animal which can move over much greater distances. Phoresy is quite common among soil animals (Binns 1982). Millipedes often carry mites on their body surface which

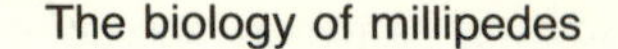

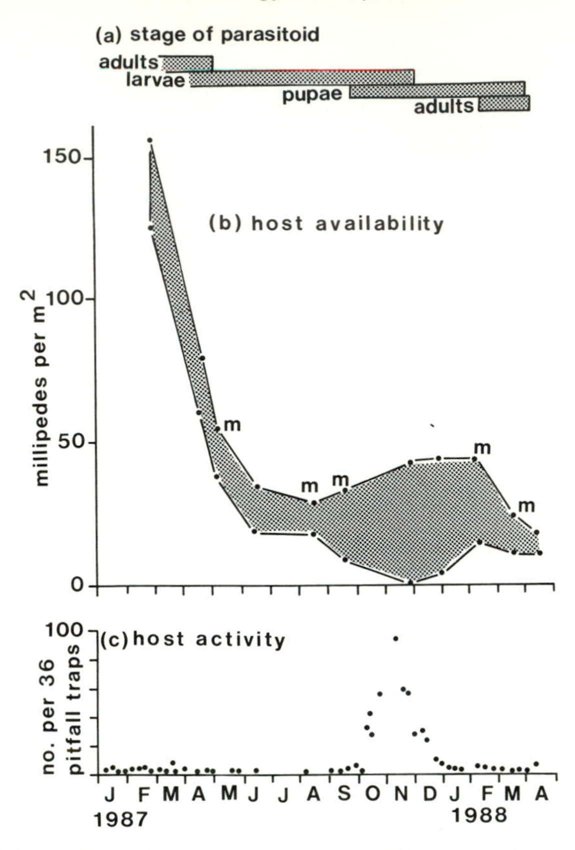

Fig. 9.5 (a) Life cycle of the dipteran parasite *Pelidnoptera nigripennis* related to (b) availability of large (> stadium VI) specimens of the millipede *Ommatoiulus moreleti* (shaded area). The unshaded area below the lower curve represents millipedes too small to be parasitized (m = moulting millipedes present in samples). (c) Surface activity of millipedes as measured by pitfall traps. Redrawn from Bailey (1989) by kind permission of the author and CAB International Institute of Entomology.

seem to do them little or no harm (Andre 1943; Fain 1987*a,b*, 1988; Kethley 1974).

The mites are usually 'resting' stages which attach themselves to the host, moult to an adult form, and then fall off. For example, deuteronymphs of *Histiostoma feroniarum* are frequently found attached to *Ommatoiulus moreleti* (Table 9.1; Baker 1985*a*).

Snider (1984*b*) examined 2000 *Polydesmus inconstans* for external parasites and found that only 54 of the millipedes had mites attached. The great majority of these mites were on the legs. The maximum number of mites on a single millipede was 106.

Table 9.1 Incidence of mites on different stadia of *Ommatoiulus moreleti* collected from Cascais, Portugal between January 1981 and January 1982. Reproduced from Baker (1985*a*) by kind permission of the author and CSIRO Australia.

O. moreleti		Numbers of Acari per millipede				Percentage of *O. moreleti*
Stadium	*N*	0	1–10	11–20	>20	with Acari
IV	16	16				0
V	137	136	1			0.7
VI	122	115	7			5.7
VII	173	158	15			8.7
VIII	317	262	55			17.4
IX	277	155	107	8	7	44.0
X	264	109	122	28	5	58.7
XI	114	21	66	16	11	81.6
XII	37	4	17	12	4	89.2
XIII	7	1	1	3	2	
XIV	1		1			

9.3 Defence

9.3.1 Introduction

Millipedes, being generally slow-moving animals, are unable to resort to running away to avoid attack, unlike their myriapod relatives, the lithobiid centipedes. One species of *Diopsiulus* (Nematophora: Stemmiuloidea) from Sierra Leone is able to jump for two to three cm (Fig. 3.15; Evans and Blower 1973) but in general, millipedes have resorted to physical or chemical means of defence.

The exoskeleton offers some protection. Some species that survive unmolested among ants may be impregnated with 'ant-like odours' (Akre and Rettenmeyer 1968; Rettenmeyer 1962; Donisthorpe 1927). However, even heavily calcified species such as *Sphaerotherium* which are protected from attack by ants, birds, and mice, are unable to resist a concerted attack from a mongoose (Fig. 9.1; Eisner and Davis 1967). Pill millipedes, such as *Glomeris marginata*, have powerful muscles that enable them to roll into a complete, tightly-enclosed sphere (Candia Carnevali and Valvassori 1982) but they may also produce defensive secretions (see below).

Defence glands are present in all millipedes except for Penicillata, Sphaerotheriida, and Chordeumatida (Eisner *et al.* 1978). There is usually only one pair per segment. This indicates that they evolved after the development of diplosegments. In most millipedes, the openings of the glands are placed laterally. In *Chelojulus sculpturatus*, the gland openings are born on lateral swellings (Enghoff 1982*b*). In *Glomeris marginata*, the

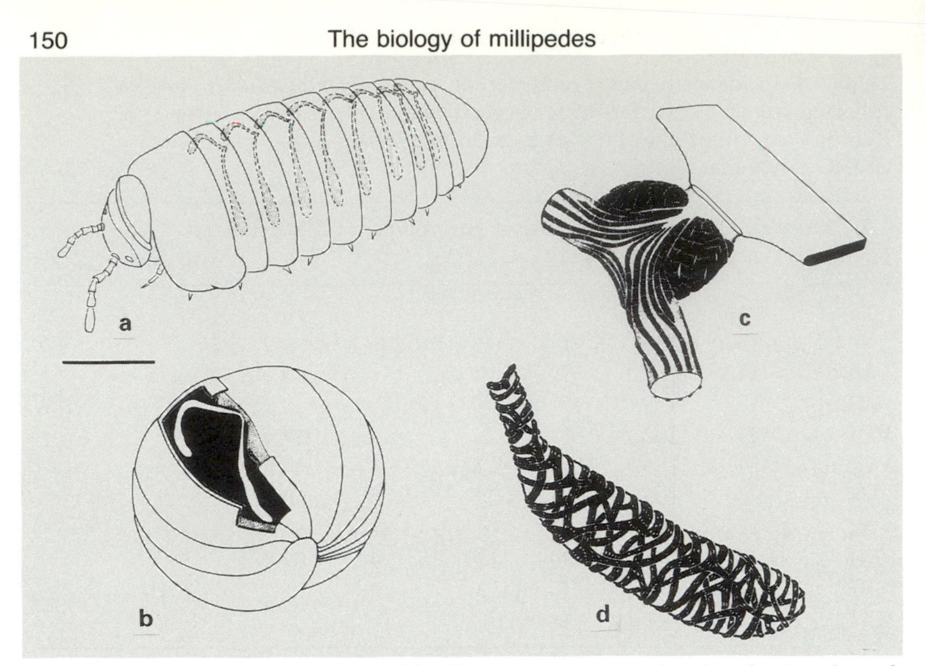

Fig. 9.6 Defensive gland Type 1. (a) *Glomeris marginata* showing the location of the eight pairs of segmental glands. Scale bar = 2 mm. (b) Coiled *Glomeris* showing cutaway with one gland pair in place. (c) Enlarged view of muscled junction of a gland pair at the level of the slit-like gland opening. (d) Portion of a glandular sac showing arrangement of investing musculature. Reproduced from Eisner *et al* (1978) by kind permission of the authors and Springer-Verlag.

paired glands on each segment open via mid-dorsally placed pores that produce single drops of secretion (Figs 9.6, 9.7(b)).

Reviews of millipede defensive secretions have been provided by Eisner (1970), Eisner and Meinwald (1966), Casnati *et al*. (1963), and Witz (1990). The excellent comprehensive review by Eisner *et al*. (1978) provides much more detail than it is possible to include here and this should be consulted for details of the literature prior to 1978.

9.3.2 Effects of defensive secretions on humans

While there is no evidence that any human has ever been killed by a millipede (they have been known to kill lizards—Stebbins 1944), their secretion may cause considerable discomfort if they come into contact with sensitive skin, and blindness if they get into the eyes (Burtt 1947; Halstead and Ryckman 1949; Haneveld 1958; Radford 1975; Smith 1973). It has even been reported that the secretion of *Spirobolus bungii* inhibits division of human cancer cells in culture (Tinliang *et al*. 1981). Some jungle tribes use millipede extracts as poison on arrowheads.

The secretion can be sprayed for a considerable distance, over 40 cm in the spirobolid *Metiche* (=*Epibolus*) *tanganyicense* (Wood *et al.* 1975).

9.3.3 Types of defensive glands

Eisner *et al.* (1978) recognized three main types of defensive glands in millipedes.

9.3.3.1 Type 1 defensive glands

Type 1 glands occur in the Glomerida and have been studied extensively in *Glomeris marginata*. This pill millipede has eight pairs of glands which open mid-dorsally on segments 4–11 (Fig. 9.6). Each gland is surrounded by a network of muscles that force secretion out through the openings. Individual specimens of *Glomeris marginata* invest between about 1 and 4 per cent of their body weight in this secretion (Carrel 1979). After an attack in the field, the glands take about four months to replenish their contents fully (Carrel 1984).

The secretion is issued from each gland as a sticky droplet (Fig. 9.7b) and contains the quinazolinones glomerin and homoglomerin. These act as anti-feedants, sedatives, and toxins to predators (Carrel 1990; Meinwald *et al.* 1966; Schildknecht and Wenneis 1967; Schildknecht *et al.* 1966). Wolf spiders are particularly sensitive to the sedative effects of the secretion (Carrel and Eisner 1984) (Table 9.2; Fig. 9.7(c,d))

The sedatives are interesting in that their chemical structure is very similar to that of the synthetic human sedative quaaludes (Harborne 1988). This is a nice example of how millipedes evolved sedatives many millions of years before their development by humans. Interestingly, the plant-derived sedative arborine, and the synthetic methaqualone do not sedate wolf spiders (Carrel *et al.* 1985).

9.3.3.2 Type 2 defensive glands

The structure of this type of gland is relatively simple (Woodring and Blum 1965). It consists of a spherical sac, containing intrinsic secretory cells, which opens onto the lateral surface of the millipede via a small pore (Fig. 9.8). The pore is held shut under normal circumstances but when discharge takes place, a muscle attached to the wall of the duct contracts, opening the duct and allowing secretion to escape. The sac is apparently not surrounded by muscles and the force to emit the secretion is probably provided by increased haemolyph pressure.

This type of gland is present in most millipede orders including Spirobolida, Spirostreptida, Julida, Callipodida, Platydesmida, and Polyzoniida (Eisner *et al.* 1978). It may also occur in Stemmiulida and Siphonophorida but there is at present no information on the structure of defensive glands in these groups.

The secretions from these glands contain a very wide range of chemicals,

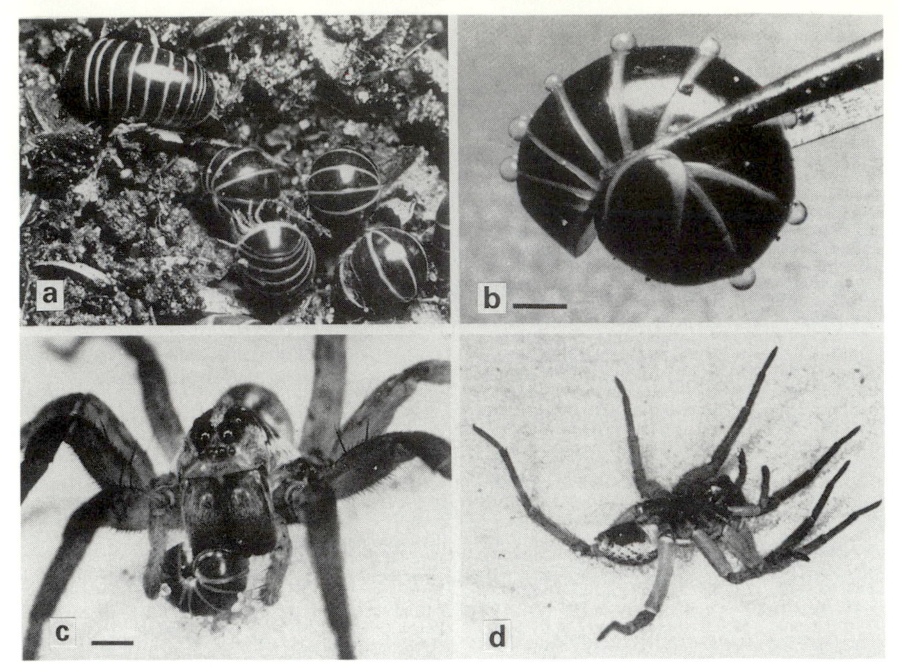

Fig. 9.7 (a) *Glomeris marginata* on soil. (b) Coiled *Glomeris* responding to slight pinching with forceps by emitting secretion from the eight mid-dorsal openings of its defensive glands. Scale bar = 1 mm. (c) *Lycosa* sp. attacking a *Glomeris* which is held, already injured, in the spider's chelicerae. Scale bar = 2 mm. (d) Spider in state of sedation, hours after having attacked a *Glomeris*. The spider has been flipped on its back with a glass rod and is failing to right itself, thus meeting the experimental criterion for sedation. Reproduced from Carrel and Eisner (1984) by kind permission of the authors.

Table 9.2 Course and outcome of attacks by spiders on *Glomeris marginata* (n = 83). Reproduced from Carrel and Eisner (1984) by kind permission of the authors.

Course of attack	Number of millipedes that discharged secretion	Condition of spiders (12 h after attack)	
		Normal, no.	Sedated, no.
Millipede inspected, released unharmed (n = 33)	0	33	0
Millipede bitten, released unharmed (in <3 min) (n = 36)	28	34	2
Millipede lethally bitten, rejected uneaten or partially eaten (in 3.5–56 min) (n = 14)	13	3	11

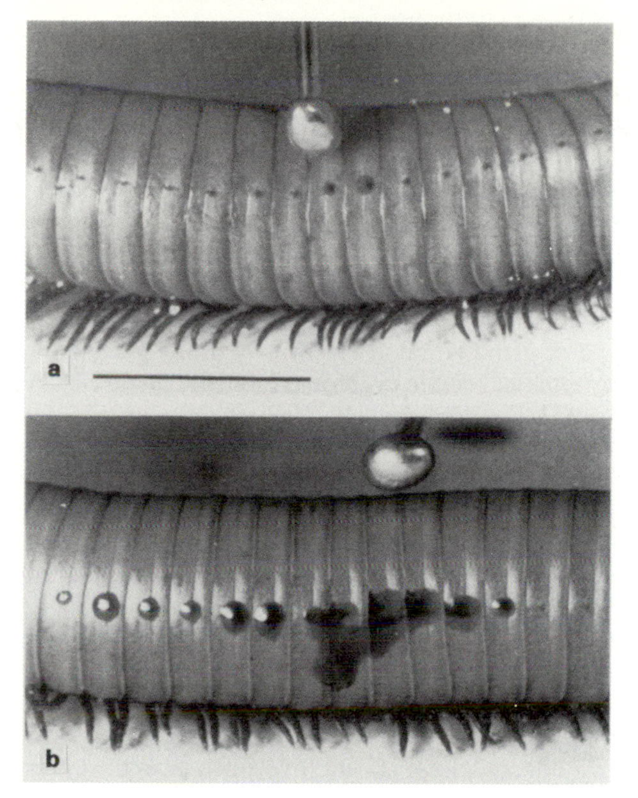

Fig. 9.8 Defensive gland Type 2. *Narceus gordanus* discharging quinonoid secretion in response to tapping with a metal mallet. The response is initially localized (a), and spreads to adjacent segments (b) after persistent stimulation. Scale bar = 1 cm. Reproduced from Eisner *et al.* (1978) by kind permission of the authors and Springer-Verlag.

mostly with low molecular weights (<300). The most common are benzoquinones which are found only in Spirobolida, Spirostreptida, and Julida. Eisner *et al.* (1978) provided a detailed list. Even the secretions of species that are apparently closely-related may show considerable differences in their chemistry (Bernardi *et al.* 1982; Röper and Heyns 1977). More systematic analysis of the precise chemical nature of the secretions could provide interesting taxonomic information.

In *Uroblaniulus canadenisi* (Julida), the secretion consists of a mixture of benzoquinones and aliphatic compounds (Weatherston and Percy 1969). In *Narceus annularis* (Spirobolida), toluquinone and 2-methoxy-3-methylbenzoquinone are the main components (Percy and Weatherston 1971). The secretion of *Rhinocricus insulatus* (Spirobolida) contains the aldehyde trans-2-dodecenal, otherwise known only from plants (Wheeler *et al.* 1964) whereas that of *Metiche* (=*Epibolus*) *tanganyicense* contains ubiquinone-0 (Wood *et al.* 1975).

Millipedes of the genus *Polyzonium* produce a nitrogen-containing terpenoid in their secretion (Smolanoff *et al.* 1975; Meinwald *et al.* 1975). They smell distinctly of camphor when irritated (Cook 1900). The secretion acts as an irritant to insects. A solution of this substance 'polyzonimine' applied topically to cockroaches induces scratching at a concentration of only 10^{-4} M (Smolanoff *et al.* 1975).

9.3.3.3 Type 3 defensive glands

The third type of gland is found exclusively in the Polydesmida, which is the only order able to secrete cyanide (Conner *et al.* 1977; Coolidge 1909; Duffey *et al.* 1977; Duffield *et al.* 1974; Eisner *et al.* 1967; Wheeler 1890). The almond-like aroma of polydesmid millipedes is familiar to diplopodologists who collect species of this group. Care has to be taken not to mix polydesmids with other arthropods in collecting tubes if all the specimens are wanted alive at the end of a day's fieldwork! Polydesmids themselves, not surprisingly, are highly resistant to cyanide poisoning (Hall *et al.* 1969, 1971).

In *Apheloria corrugata*, the secretion is emitted from glands which each contain two compartments (Fig. 9.9a,b; Eisner *et al.* 1963). The inner compartment stores mandelonitrile while the outer (smaller) compartment (from which it is separated by a valve) contains an enzyme that catalyses the breakdown of mandelonitrile into hydrogen cyanide and benzaldehyde (Fig. 9.9c). The contents of these chambers are produced by specialized secretory cells either in the walls, or connected to them by short cuticle-lined ducts. The synthetic pathway for mandelonitrile is similar to that of plants (Duffey *et al.* 1974). All the available evidence suggests that millipedes manufacture these substances from simple precursors, rather than obtain them from their plant food or symbiotic microorganisms (Duffey and Blum 1977; Towers *et al.* 1972).

The glands are not present on every ring and in *Apheloria corrugata*, they are present only on segments 5, 7, 9, 10, 12, 13, and 15–19. In addition to their defensive properties, the secretions may also protect against fungal attack. The secretions of *Oxidus gracilis* were able to suppress mycelial growth and spore germination in ten species of fungi isolated from millipedes and adjacent soil (Roncadori *et al.* 1985).

Polydesmids are highly effective at repelling predators with their secretion. *Pachydesmus crassicutis* was able to repel fire ants for some time before its secretion was exhausted (Blum and Woodring 1962). It is interesting to note that recent work by Peterson (1986) has shown that the ant *Myrmica americana* does not find hydrogen cyanide repellant and is deterred from attacking polydesmids by other components of their secretion (Table 9.3). Perhaps the benzaldehyde serves a 'signal function' to

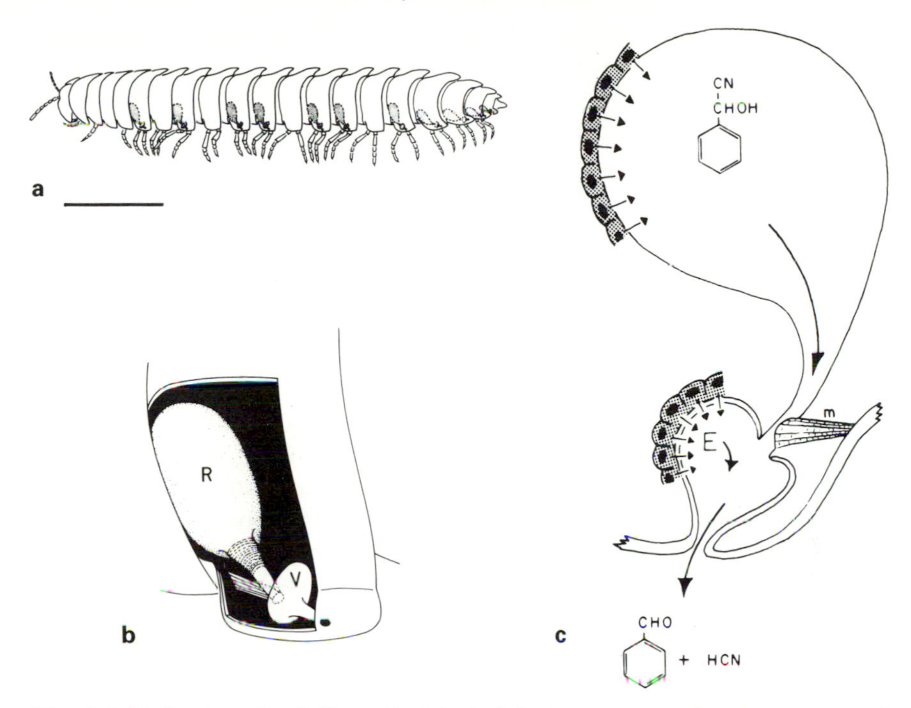

Fig. 9.9 Defensive gland Type 3. (a) *Apheloria corrugata* showing segmental arrangement of the glands. Scale bar = 1 cm. (b) Enlarged cutaway view of gland. R, reservoir; V, vestibule. (c) Diagram of the two-chambered cyanogenetic glandular apparatus of *Apheloria corrugata*. The inner compartment (reservoir) stores mandelonitrile while the smaller compartment (vestibule) contains an enzyme, E, that catalyses the breakdown of mandelonitrile into hydrogen cyanide and benzaldehyde. A muscle, m, operates the valve between the two compartments. Reproduced from Eisner *et al*. (1978) by kind permission of the authors and Springer-Verlag.

repel predators from cyanogenic prey? Alternatively, the ants may have evolved to tolerate it.

9.3.4 Effectiveness of defensive secretions on predators

The topic of predation has been covered in Section 9.1. However, it is worth returning to this subject briefly in the light of the above descriptions of the defensive glands. Many experiments have been conducted in the laboratory which have shown that the defensive secretions are effective against a wide range of predators (for a review see Eisner *et al*. 1978). However, predation in the field has rarely been observed.

Millipedes will generally only discharge their secretion from threatened segments (Fig. 9.8). Thus, they are protected from repeated attacks by small predators such as ants, providing they do not overwhelm the millipede

Table 9.3 Behavioural responses of ants to vapours of test solutions. Treatments followed by different letters were significantly different at $P < 0.05$. Reproduced from Peterson (1986) by kind permission of the author and Springer-Verlag.

Test solution	Degree of response		
	1 = no effect	2 = brief pause	3 = withdrawal
Phosphate buffer (a)	40	3	2
HCN (a)	41	2	2
Benzaldehyde (b)	3	11	31
HCN + benzaldehyde (b)	5	9	31

by sheer force of numbers. However, there is no doubt that large numbers of millipedes fall victim to predation and that some invertebrates (including several species of ants) do not appear to find the secretions repellant at all.

We should then ask, when did defensive glands evolve, and what were the predators which stimulated their development? Very few temperate millipedes possess warning coloration, which is surprising considering their horrible taste. This would suggest that birds are/were not major predators of temperate millipedes. In the tropics, however, there are many species that are very brightly coloured (see Section 3.2).

Unfortunately, there is little fossil evidence which might help us in elucidating the evolution of defensive glands. However, it was probably the arrival of the ants which provided the selective pressure for the evolution of the glands. For some species, the secretion still provides an effective defence but for others, the ants have 'out-evolved' the millipedes and are able to prey on them. Further evolutionary pressure may have been provided with the arrival of small mammals. The ability of some tropical species to forcibly eject their secretions over considerable distances (over 40 cm) is unlikely to have evolved in response to invertebrate predators.

Further studies are required on predator–prey relationships in millipedes. It would be particularly interesting to examine predation of introduced species by native animals.

9.3.5 Bioluminescence

While the cause of bioluminescence in millipedes is unknown, it is likely that it acts as a warning. Causey and Tiemann (1969) listed several potential predators and also recorded that on a dark night, the millipedes could be seen more than 10 metres away. They were sometimes present in such large numbers that they resembled the 'starry sky on a dark night'. It seems that being night active, these millipedes are advertising their unpleasant taste.

Nearly all the species of the genus *Motyxia* (= *Luminodesmus*) (Polydesmidae) in the USA are luminescent. The genus is centred in California and Causey and Tiemann (1969) have suggested that the species may be derived from a single ancestor.

The most studied species is *M. sequoiae*. This lives in moist humus, is approximately 40 mm in length, and is pale tan in colour with a darker dorsal longitudinal stripe. In the dark, the whole animal including its appendages glows a bright greenish-white. Davenport *et al.* (1952) followed the development of luminescence along with that of the animal. The eggs were not luminous. In the first stadium a faint glow could be seen when the animals were in a mass. In the second stadium, the millipedes were obviously luminous.

The glow is brighter from thicker areas of the integument and is brightest in 100 per cent oxygen, at 31.5 °C. It is continuous but fluctuates visibly by 20–40 per cent. The intensity can increase instantaneously, for example when the animal is handled. The emission spectrum has its maximum in the green at 495 nm which is similar to that of bioluminescent bacteria. However, Hastings and Davenport (1957) concluded that bacteria were not responsible. Pieces of the millipede, such as the legs or decapitated body, retain luminescence with a half-life of about eight hours. Causey and Tiemann (1969) thought the cause to be a secretion from the repugnatorial gland. It is possible to produce an extract which is luminous for a short time. *M. sequoiae* are also fluorescent, as are the other species in the same genus.

10

Ecology

10.1 Introduction

Many of the earlier workers, such as Schubart and Verhoeff, included valuable ecological details in their treatises on identification. However, many of their comments were inferred from preserved millipedes. More recent identification guides (e.g. Blower 1985; Demange 1981) have included ecological information based on field and laboratory studies of living animals.

If we consider millipede ecology in the broadest sense and include details of life cycles, development, population dynamics, role in decomposition, status as pests etc., there is an abundance of information. Ongoing ecological studies are being carried out in several countries ranging from France and Greece to Eastern Europe and the Soviet Union, the Ivory Coast, and India. From this work, certain general principles have emerged. These are reviewed in this chapter.

10.2 The role of millipedes in decomposition processes

The impact of millipedes on soil processes varies according to the species and the characteristics of the site concerned. It is evident that in some places their impact is of fundamental importance, whereas in others they fulfil a very minor role. Their main effect is probably one of fragmentation which stimulates microbial activity and indirectly influences the fluxes of nutrients (Fig. 10.1; Anderson and Leonard 1988; Anderson *et al.* 1985).

The fragmentation of leaves, stimulation of microbial activity, and subsequent depositon of faecal pellets has important ecological implications (Hanlon 1981*b*). Indeed, although millipedes (and other saprophages) are not responsible directly for more than 10 per cent of chemical decomposition, their feeding activities are vital in stimulating the microorganisms which carry out some 90 per cent of chemical breakdown (Anderson and Bignell 1980; Van der Drift 1951). Furthermore, species which feed on the surface during the night will promote decomposition further by innoculating leaf litter with fungal spores and bacteria in the deep, moister layers when they move down during the day. This may be particularly important in desert environments (Taylor 1982*a*). A lack of millipedes (and earth-

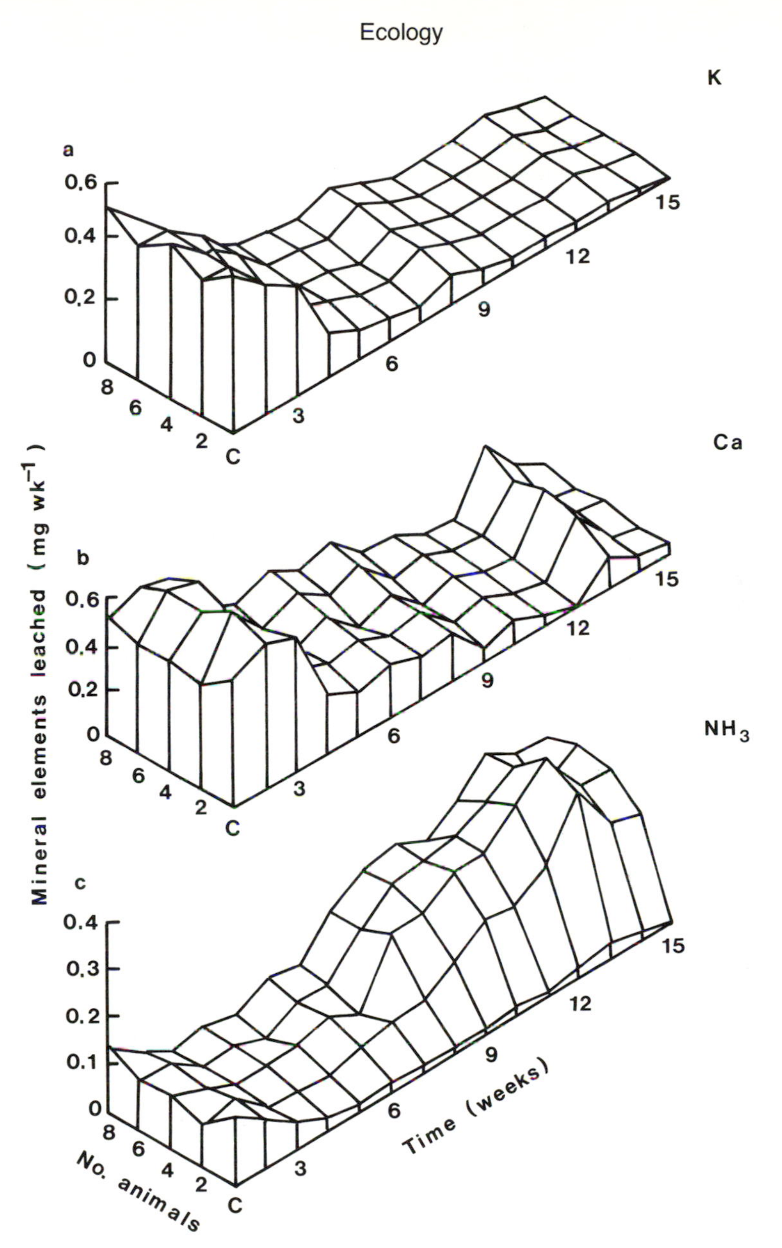

Fig. 10.1 Mobilization of (a) potassium, (b) calcium, and (c) ammonium–N by different numbers of *Glomeris marginata*. Animals were added at week 3 to all treatments except controls. Note that the leaching rates of ammonium–N (c) are increased significantly in the presence of millipedes in comparison to controls. Redrawn from Anderson *et al*. (1983) by kind permission of the authors and Pergamon Press.

worms) is thought to be one of the main reasons for the slow decomposition rate of pine needles.

Diplopods are certainly good fragmenters and consume much leaf litter. Gilyarov (1970) quoted figures of 80–90 per cent of ingested matter voided as excrement (see also Chapter 4). According to Jackson and Raw (1966), 'in places where they are abundant, as in some woodland soils, a layer consisting largely of millipede faecal pellets may occur'. In the pill millipede *Glomeris marginata*, for example, these pellets contain leaf fragments of 0.001–0.1 mm in diameter and a mixture of spores, pollen, fungal mycelia, algae, and bacteria (Nicholson *et al.* 1966).

Whilst Lyford (1943) considered that millipedes ate only between 1 and 5 per cent of the leaf litter that falls to the forest floor, Striganova (1970) has estimated that in the Caucasus, millipedes can consume nearly all the leaf litter. David (1987*a*) estimated that adult *Allajulus (Cylindroiulus) nitidus* consume 3–4 per cent of leaf litter falling each year in forests in Orleans, France. However, this species represents only a part of the millipede fauna and taking into account other species and juveniles, probably between 9 and 16 per cent of the leaf fall is eaten by millipedes. In a French oak forest, Bertrand *et al.* (1987) estimated that *Glomeris marginata* consumed 8–11 per cent of the annual leaf fall.

10.3 Population densities

Numerous estimates have been published of the densities of millipedes in a wide range of habitats. Kime (1991) quoted 0–1000 m^{-2}, although densities higher than this can occur in 'swarms' (Section 10.5; Fig. 1.3). In a survey by Iatrou and Stamou (1989) of macroarthropods in a Mediterranean ecosystem, millipedes were found to be the most numerous saprophage. They occurred at a density of 114 m^{-2} (total arthropod population density 316 m^{-2}), i.e. 38 per cent of the total. Fig. 10.2 gives an indication of the position of millipedes in a French forest where their relative importance is more typical of temperate woodland.

Some within-site variability in density may be due to differences in sampling methodology (Geoffroy 1981*a*), as well as time of year. Striganova and Rachmanov (1972) recorded a peak density of *Amblyiulus continentalis* in a broadleaf Caucasian woodland of 45–65 m^{-2} in autumn which dropped to only 1–2 m^{-2} in summer. Gilyarov (1979) recorded fewer millipedes under fir than beech. Aouti (1978) found fewer millipedes in a plantation of *Hevia brasiliensis* in the Ivory Coast (8000 ha^{-1}) than in a teak forest (20 000 ha^{-1}). David (1987*b*) noted that there is a relationship between the size of the millipede population and the quality of the humus. The better the quality, the greater the population and consequently, the shallower the litter layer.

Some of the highest population densities have been recorded for *Pro-*

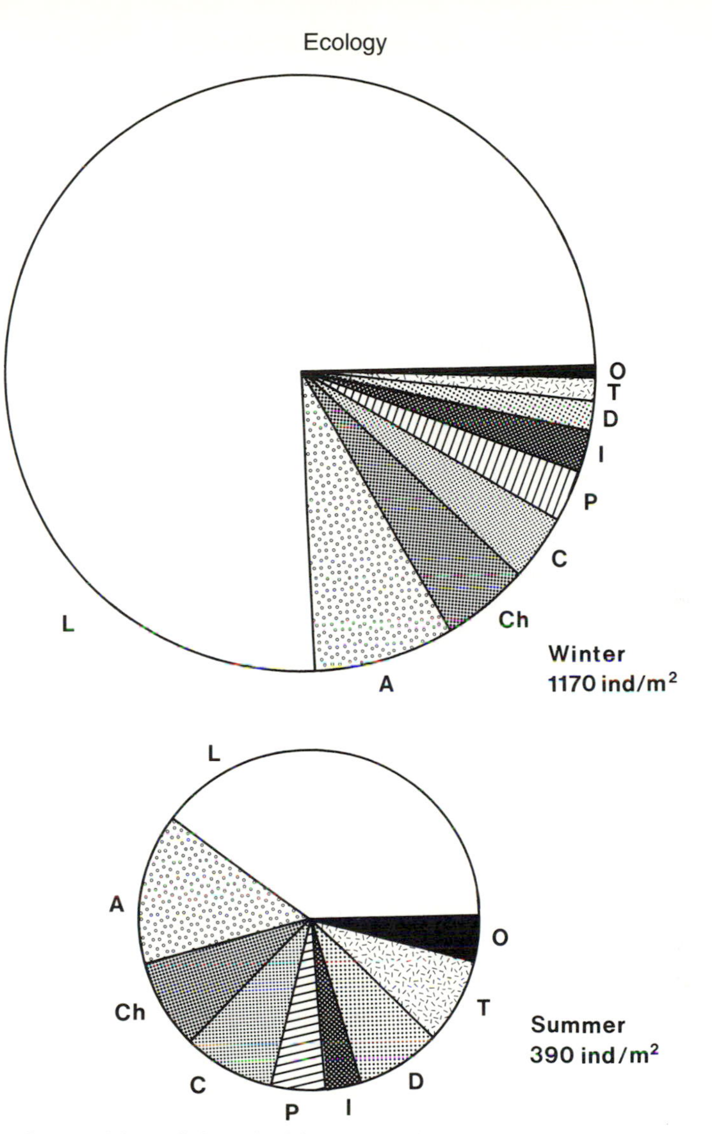

Fig. 10.2 Composition of the edaphic macroarthropod populations in winter and summer in a mixed forest in France. A, Arachnida; C, Coleoptera; Ch, Chilopoda; D, Diplopoda; I, Isopoda; L, endopterygote larvae; O, other macroarthropods; P, pseudoscorpions; T, Thysanoptera. Redrawn from Geoffroy (1981a) by kind permission of the author.

teroiulus fuscus. Elliott (1970) recorded 10^6 juveniles per acre and Tracz (1987) counted 2000–3000 m^{-2}. Critchley *et al.* (1979) pointed out that in ecological terms, it is biomass which is more important than the absolute numbers (although biomass figures are often more difficult to obtain).

Millipedes frequently have an aggregated distribution. This may be obvious as when turning over a log reveals a polydesmid family group, or in arid conditions where millipedes collect under a stone at the base of a bush. However, aggregations occur also in more uniform habitats such as leaf litter. Aggregations may vary with depth of soil (Bandyopadhyay and Mukhopadhyaya 1988) and are often related to humidity (Peitsalmi 1974). The sex ratios of aggregations may also be variable (Bano and Krishnamoorthy 1979).

Bhakat (1987) estimated the density and biomass of *Streptogonus phipsoni* (Polydesmida) in grassland over a year by sampling with quadrats (Fig. 10.3). There were considerable differences between months. In general, the mean population density for each month was correlated positively with mean monthly rainfall.

The studies of David in France have dealt mainly with populations of diplopods and their effects on forests soils. In one such study, David (1984) considered just one species, the chordeumatid *Microchordeuma* (=*Melagona*) *gallicum*. The maximum density in May corresponded with the recruitments from the new generation, the minimum in February/March represented the end of the old generation.

A smaller number of authors have produced life tables which include estimates of production. Notable amongst these is Crawford (1976) who calculated annual production of a desert species *Orthoporus ornatus*. A figure of 0.85 kg ha^{-1} was obtained with ingestion estimated at 3.434 kg ha^{-1} and defecation at 3.181 kg ha^{-1}. However, it should be noted that in these extreme conditions, production was limited to three to four months of the year.

The most comprehensive study to deal with density, production, and related factors, was that of Blower (1970*a*). This was notable not only for its thoroughness, but also for the inclusion of information on all the different species at the woodland site studied. *Julus scandinavius* was studied in detail with regard to standing crop and production, and each stadium was considered separately. Life table details such as fecundity and survival were also calculated (see also Sections 8.5. 3.3).

10.4 Millipedes as pests

In spite of the important role that millipedes play in natural decomposition processes in terrestrial ecosystems, all articles that deal with the economic importance of millipedes lay emphasis on their status as pests. As early as the mid nineteenth century, Curtis (1860) included a section on millipedes in his book *Farm Insects* (*sic*). Delightful hand-coloured illustrations were included for several species including the spotted snake millipede *Blaniulus guttulatus*, and a variety of polydesmids and julids.

Brade-Birks (1929, 1930) commented that while centipedes are beneficial, millipedes are detrimental, and this seems to be the consensus of

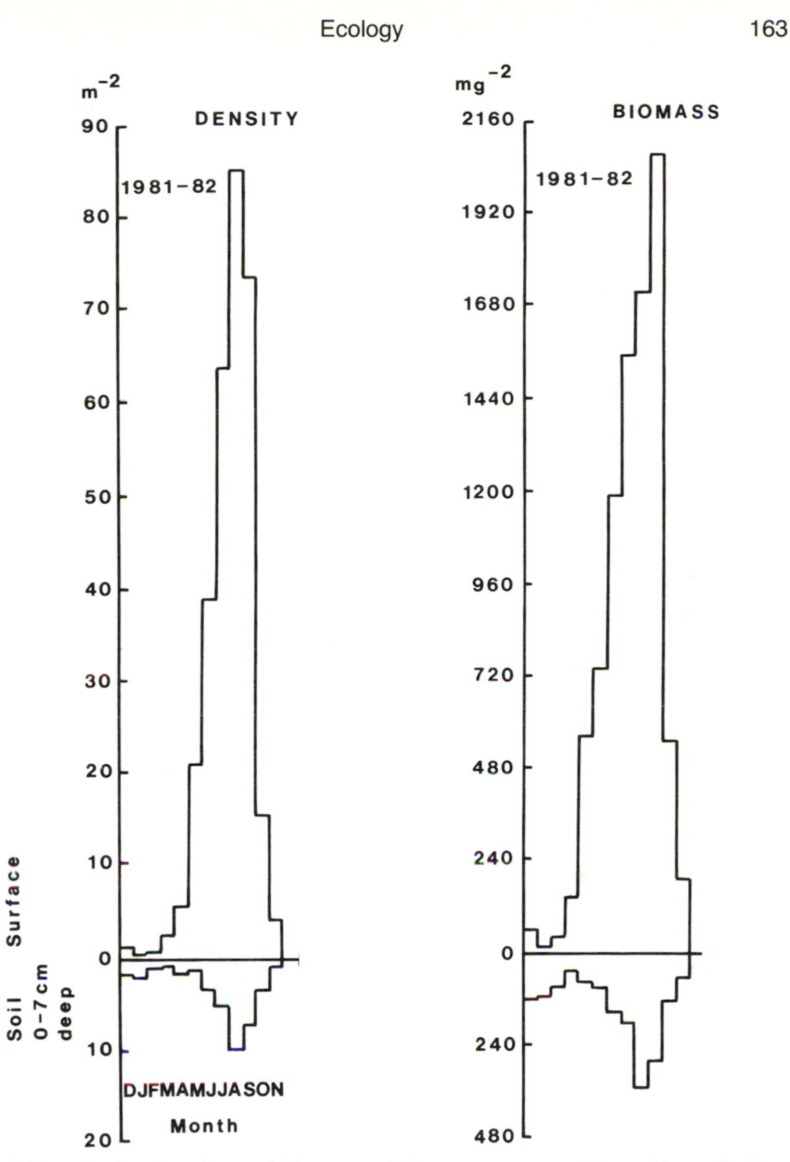

Fig. 10.3 Population density and biomass of *Streptogonopus phipsoni* in an Indian grassland from December 1981 to November 1982. Redrawn from Bhakat (1987) by kind permission of the author and the Zoological Society of London.

opinion among those concerned with crop plants. There is no doubt that millipedes have at times been labelled justifiably as pests. However, compared with insects, their economic effects on a global scale are negligible. The type of nuisance may be manifested in two ways. First as feeders on crops, and second, when they occur in large aggregations, especially in houses.

There have been various claims made of millipedes causing damage to crops. Most frequent reports are for sugar beet and potatoes, but strawberries, cucumbers, orchard fruit, and roots of wheat and flowers have been reported as targets for *Brachydesmus superus, Oxidus gracilis*, and various blaniulids (Julida) in temperate regions (MAFF 1984). In Africa, cotton and groundnuts are eaten by spirostreptid millipedes. Numbers at times can reach epidemic proportions. Densities of *Oxidus gracilis*, for example, can exceed 2500 m^{-2} in some heated glasshouses (Edwards and Gunn 1961).

Cloudsley-Thompson (1950) stated that between 1944 and 1949, more than 100 cases of millipede damage were recorded by advisory entomologists in Great Britain; 90 per cent of these concerned the 'snake spotted' millipede *Blaniulus guttulatus* and were mostly on potatoes or sugar beet. According to Biernaux (1966), millipede damage to a sugar beet field in Belgium in 1965 caused a 22 per cent decrease in total crop weight and a 25 per cent decrease in weight of sugar subsequently produced. Thus millipedes can be of considerable local economic importance.

It has been suggested that millipedes do not cause the primary damage in tubers, but come in after the skin has been broken (Brade-Birks 1923). Vachon recorded 120 blaniulids from a single potato whilst the rest of the crop was undamaged (cited in Cloudsley-Thompson 1950). In contrast, Pierrard and Biernaux (1974) noted that *Blaniulus guttulatus* causes primary damage, and then polydesmids and other julids follow on. Many millipedes are certainly attracted to cut surfaces. Some may attack crops as a source of water during dry spells (Brade-Birks 1923).

Millipedes attack the tap roots of many plants but they may also cause damage to seedlings. According to Biernaux (1966), a cold spring prevents quick growth of the young plants and up until the six to eight leaf stage, they are vulnerable to attack. Symptoms are wilting and nutrient deficiency. Fields with high soil organic matter content are more likely to be affected.

Various methods of dealing with the pests have been tried. Cloudsley-Thompson (1950) recommended burning the soil rather than removing the humus. Appel (1988) demonstrated the success of integrated research. By investigating the humidity preferences of *Oxidus gracilis* in Alabama, he recommended the removal of all excess damp organic matter and debris from gardens and associated areas and achieved a decrease in the numbers of millipedes found inside buildings from 200 to 15 per day.

A variety of chemicals have been tested on millipedes with contradictory results. For example, Fleming and Hawley (1950) found that DDT had no effect. Gilyarov (1970) on the other hand reduced the population of millipedes by a factor of 19 by application of DDT at 30 kg ha^{-1}. Edwards (1974) found that such differences were due, not surprisingly, to the dose of DDT applied.

Most organophosphate insecticides seem to have little or no appreciable effect on millipedes (Edwards 1974; Pierrard and Biernaux 1974). Heptachlore was found to have a high effect which was persistent (Biernaux *et al.* 1973). Extensive tests by Edwards and Gunn (1961) showed that Lindane and Dieldrin produced low initial mortality in *Oxidus gracilis*, but after one month, infestations had declined to a low level. In the same study, other pesticides were found which might produce a knock-down effect, but numbers of millipedes quickly rose again. Biernaux (1966) reported that Dieldrin was not effective for the control of julids.

Differing effects on different species of millipedes were also found in California by Rust and Reierson (1977). Methonyl, Carbaryl, and Propuxur were effective for dealing with *Oxidus gracilis*, but not *Bollmaniulus* sp. Carbaryl seems to be one of the most effective chemicals for controlling migrations of millipedes. It is used as a barrier to prevent the animals from crossing from one area to another and has been used widely in this way in Australia.

Pierrard and Biernaux (1974) recommended seed dressing for the protection of cotton, but, as this cannot be used for ground nuts, they laid a bait of an attractive substance mixed with a toxic one. This was highly effective in killing millipedes in Senegal.

Jolivet (1986), despite trying many different chemicals, could never obtain 100 per cent mortality. His study concerned an unidentified spirostreptid millipede of the genus *Spinotarsus* that had been introduced to the Cape Verde Islands. The presence of this millipede has severely restricted the growth of potatoes on the islands. Biological control has been considered but has been hindered by lack of knowledge of the biology of the species and its country of origin. Research is underway to try to discover the native habitat of the species and, hopefully, some associated parasite that might aid in its control.

Ommatoiulus moreleti is another introduced species which has become a pest, mainly in Australia. Because of this, it has become one of the most extensively studied millipedes. *Ommatoiulus moreleti* is a julid millipede, formerly indigenous to the Iberian Peninsula, which has become well established in many parts of the world (Baker 1984). It was first recorded in Australia at Port Lincoln in 1953 (Baker 1985*c*). Since that time, it has spread widely and become a considerable pest. Besides eating garden vegetables, it invades houses, where large numbers of the millipedes are squashed. As Baker (1978*c*) has recorded, 'in short, they are a revolting nuisance'.

Baker and colleagues have followed the progress of *Ommatoiulus moreleti* in Australia from the mid 1970s to the present. A substantial program of research has been carried out to explore the feasibility of biological control. The spread of the species from Bridgewater, South Australia, was mapped by Baker (1978*c*) from information on the locations of applications for licences for pesticides to deal with the invasions between 1964 and

1972. The rate of expansion was approximately 100 to 200 metres per year and was much the same in built-up areas and in pastureland. Dispersion also appeared to occur along major roads and railways.

In 1978, there was no evident decrease in the number of millipedes near the origin in comparison with peripheral areas, contrary to the predictions of Caughley's (1970) model. The model predicts that as resources become strained, local enemies adapt and the numbers of the introduced species declines. The number of applications for insecticides increased from 1011 in 1974 to 2353 in 1977 (Baker 1979*a*). However by 1985, sampling of old and new sites revealed the expected distribution of numbers of animals rising to a peak and then declining as distance increased from the origin (Fig. 10.4; Baker 1985*c*). The same distribution pattern was also demonstrated for one site over a period of time. Thus, it seems that a decline in numbers did occur after the initial eruption. The pattern has since been confirmed by McKillup *et al.* (1988).

The millipedes are a particular problem during the summer and early autumn, during the breeding season (Baker 1979*b*). Invasions are also bad just after rainfall. Some invasions also occur in the spring as the weather starts to warm up.

Applying chemical control is not totally satisfactory. In Spain and Portugal, the species is not particularly common and it seems likely that this is due to much heavier predation. Both hedgehogs and the large beetle *Staphylinus*

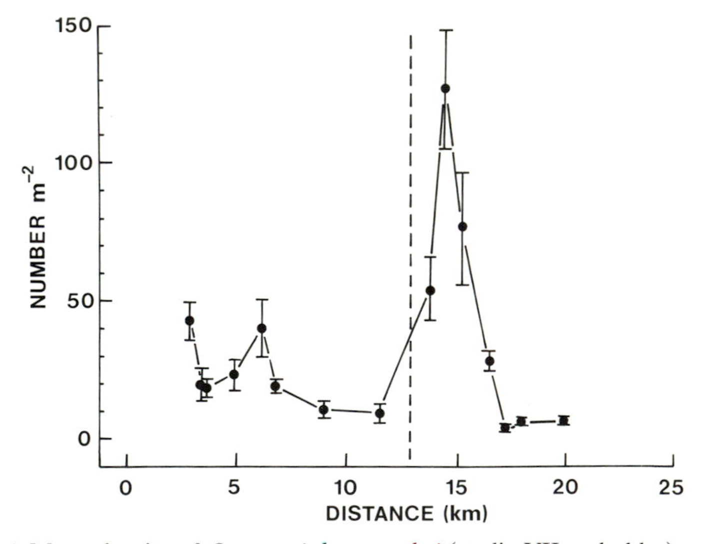

Fig. 10.4 Mean density of *Ommatoiulus moreleti* (stadia VII and older) at sites of increasing distance from Port Lincoln in October 1983. Vertical bars indicate standard error. The broken line indicates the boundary of the distribution in 1975. Redrawn from Baker (1985*c*) by kind permission of the author and Blackwell Scientific Publications.

olens were found to be very effective predators (Baker 1985*b*). However, as they are not prey-specific, more problems may be caused by introducing such animals into Australia than those that would be solved by reduction in the populations of *Ommatoiulus moreleti*. Parasites are more likely to be effective and have been the subject of a considerable research effort (see Section 9.2.8 and 9.2.9 for further discussion).

In the meantime, light traps containing insecticide such as Baygon (Fig. 10.5) can be used to lure the millipedes away from houses (McKillup 1988; McKillup and Bailey 1990; Schulte 1989*b*). *Ommatoiulus moreleti* is positively phototactic at night and invades buildings because it is attracted to the light coming from under doors.

10.5 Dispersal, mass migrations, and swarming

Millipedes have rather limited powers of dispersal. A certain amount of passive transport must take place as manifested by the world-wide distribution of some European julid species such as *Cylindroiulus latestriatus* and *Ophyiulus pilosus*. Man must in the majority of cases be the vector (see Section 2.4).

Wind has been suggested as a method of dispersal for small species by Haacker (1968) but this is probably a rare occurrence. Dispersal of millipedes takes place primarily by walking. Some species have a great tendency to wander whereas others are more or less confined to their optimum habitat. As seen in Chapter 8, the life cycles of some species allow them to colonize new areas, and/or to wander widely. In addition, some species appear to actively migrate *en masse*.

The occurrence of a large number of millipedes together at one time is one feature of their ecology which has caught the imagination of many writers. In fact, it must be one of the most widely-documented aspects of their behaviour. Despite this, the reasons for such aggregations still remain uncertain.

The numbers of animals involved can be astronomical and may be impossible to count. Cloudsely-Thompson (1949) described an incident in West Virginia USA concerning *Fontaria virginiensis* where the millipedes swarmed over 75 acres of farmland. In places, the millipedes were so numerous that the cattle would not graze, workmen were nauseated by the odour and wells filled to 15–20 cm with millipede corpses. On some days, half a barrel full of animals could be collected from one place. At night and on cloudy days, the swarm was on the move; the majority (estimated at 65 million animals) were killed by hot sunlight.

Swarms have also stopped trains. In Hungary in 1878, France in 1900, and Germany in 1906 and 1938, the rails had to be strewn with sand before the wheels would grip. In Japan, *Parafontaria laminata* has become notorious for holding up trains (Fig. 1.3; Saki 1934; Niijima and Shinohara

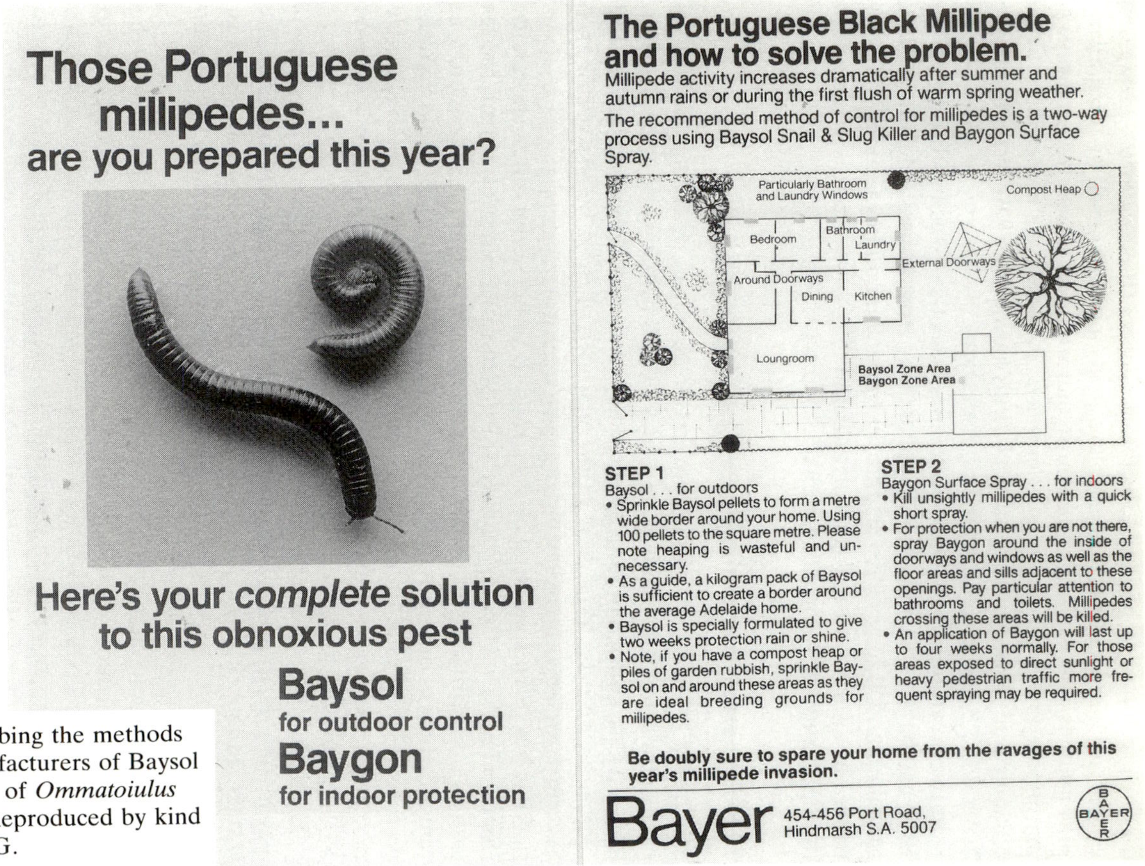

Those Portuguese millipedes... are you prepared this year?

Here's your *complete* solution to this obnoxious pest

Baysol
for outdoor control

Baygon
for indoor protection

The Portuguese Black Millipede and how to solve the problem.

Millipede activity increases dramatically after summer and autumn rains or during the first flush of warm spring weather.

The recommended method of control for millipedes is a two-way process using Baysol Snail & Slug Killer and Baygon Surface Spray.

Particularly Bathroom and Laundry Windows
Compost Heap
Bedroom
Bathroom
Laundry
External Doorways
Around Doorways
Dining
Kitchen
Loungroom
Baysol Zone Area
Baygon Zone Area

STEP 1
Baysol . . . for outdoors

- Sprinkle Baysol pellets to form a metre wide border around your home. Using 100 pellets to the square metre. Please note heaping is wasteful and unnecessary.
- As a guide, a kilogram pack of Baysol is sufficient to create a border around the average Adelaide home.
- Baysol is specially formulated to give two weeks protection rain or shine.
- Note, if you have a compost heap or piles of garden rubbish, sprinkle Baysol on and around these areas as they are ideal breeding grounds for millipedes.

STEP 2
Baygon Surface Spray . . . for indoors

- Kill unsightly millipedes with a quick short spray.
- For protection when you are not there, spray Baygon around the inside of doorways and windows as well as the floor areas and sills adjacent to these openings. Pay particular attention to bathrooms and toilets. Millipedes crossing these areas will be killed.
- An application of Baygon will last up to four weeks normally. For those areas exposed to direct sunlight or heavy pedestrian traffic more frequent spraying may be required.

Be doubly sure to spare your home from the ravages of this year's millipede invasion.

Bayer 454-456 Port Road, Hindmarsh S.A. 5007

BAYER

Fig. 10.5 Leaflet describing the methods recommended by manufacturers of Baysol and Baygon for control of *Ommatoiulus moreleti* in Australia. Reproduced by kind permission of Bayer AG.

1988). These latter outbreaks are interesting because they seem to occur at intervals of seven to eight years. Here, the majority of animals are adults plus a few sub-adults. This has led to speculation that the swarms are for mating.

Lawrence (1952) studied a swarm of *Gymnostreptus pyrrocephalus* in Natal and found it to consist of mainly adults with a sex ratio of close to one male to every two females. In contrast, Ramsey (1966) found a predominance of immatures (1 adult to 7 immatures) in a swarm of *Pseudopolydesmus serratus* in Ohio, USA. Morse (1903) recorded an adult: juvenile ratio in a swarm of *Fontaria virginiensis* of 1:300.

In addition to mating, various other explanations have been proposed for such large aggregations. Moisture appears to play an important part. Koch (1985) considered that vast swarms of *Unixenus nijobergi* (Polyxenida) were activated by rainfall. Brade-Birks (1922) also related such activity to rainy weather and proximity of vegetables and other fresh food. Population explosions in general have also been postulated as the cause (Mitra 1976) as have changes in climatic conditions or food supply (Helb 1975), and an increase in population pressure as a result of human development.

Morse (1903) recorded that each migration seemed to have its own cause and that still holds true today. However, the majority of such mass occurrences have not been studied with modern techniques. One reason for the rather anecdotal quality of most of the reports is the unpredictability of such outbreaks. Hannibal and Talerico (1986) recorded that aggregations on a farm in Kentucky occurred regularly in the spring from 1958 onwards, but in general, such phenomena are one-off events. Broadly speaking, most mass occurrences seem to occur in the spring or autumn in temperate zones or the rainy season in the tropics.

Scott (1958*a,b*) recorded millipedes that entered his house in Henley, England over a number of years. *Tachypodoiulus niger* was the most common invader. Indeed, schizophilines (including *Ommatoiulus moreleti*) have reputations as wanderers (see also Verhoef 1900), and may occur in large numbers. Other millipedes which may be very common near houses include *Oxidus gracilis*, several thousand of which were found in one house after rain in Tennessee. Most of Lenoir City was infested with the species in the same outbreak, especially houses near stream beds and woodland (O'Neill and Reichle 1970).

Many of the swarms appear to have no directional movement but there have been exceptions. One of these was described in detail by Bellairs *et al.* (1983) and concerned the Indian polydesmid *Streprogonopus phipsoni*. Large aggregations of adults of this species behave in one of four ways. First, the millipedes browse in a closely-packed group which moves a little but in no specific direction. Second, animals all move/migrate in the same specific direction, often at quite a fast speed. Third, millipedes meet an obstacle and build up into a pile. Fourth, disaggregation and reaggregation takes place, where animals spread out and then cluster in small groups which gradually amalgamate, the latter being a possible reaction to danger.

In moving aggregations of *Streptogonopus phipsoni*, individuals walk forward over the backs of others. On reaching the front, they lead briefly before being overtaken. It is possible that they are feeding at this point on cryptograms on the ground, and use the aggregation as a method of avoiding predation and reducing water loss. In a typical aggregation, it takes about 13 seconds for an animal to pass from the rear to the front and another 13 seconds to reach the back again. It then appears to browse for a further 20 seconds before hurrying to reach the aggregation. The cluster itself consists of a compact core with a few outriders and/or tails of younger animals. It moves as a whole at speeds of between 3 cm and 50 cm in an hour. On hard paths, the rate is faster and may reach 1.2 metres per hour (Bellairs *et al.* 1983).

10.6 Activity times

Most millipedes are active at night. Dondale *et al.* (1972) found that 90 per cent of the captures (in pitfall traps) were made between 22.00 and 04.00. Banerjee (1967*a*) sequentially trapped three species in pitfall traps divided into four periods of time. He found peaks in phases 1 (midnight to one hour before sunrise) and 4 (one hour after sunset to midnight).

Bano and Krishnamoorthy (1979) studied circadian rhythms in *Jonespeltis splendidus* (=*Anoplodesmus saussurei*) (Polydesmida). This species has a bimodal activity period and is crepuscular (active at dawn and dusk). When kept in the laboratory under a 12 hour light/12 hour dark regime, the activity peaks were defined very sharply with that of the morning being greater. In continuous light or total darkness, the peaks shifted erratically or did not occur at all. Vestiges of a circadian rhythm can also be detected in cave millipedes (Mead and Gilhodes 1974).

Seasonal activity is also governed by various factors. Banerjee (1967*a*) recorded that peak activity of *Cylindroiulus punctatus* occurred during the breeding season when adults are searching actively for mates.

The activity periods of some Mediterranean species are very pronounced (Karamaouna 1987; Karamaouna and Geoffroy 1985). Animals are only active in the wet period (winter and spring). No millipedes are seen between May and October when it is very dry. During this period, they burrow into the soil. Climate was also the critical factor in determining the activity times of the millipedes in Senegal studied by Gillon and Gillon (1976). Here, animals only appeared after the onset of the rains. However, activity ceased before the end of the rain in anticipation of the coming dry season.

In temperate regions, most species are less active in the winter. The main exceptions are the Chordeumatida which are adult during the winter months. Species of *Polydesmus* are active in spring and summer (occasionally autumn too) whereas those of *Cylindroiulus* usually have two regular peaks of activity in spring and autumn. By plotting temperature and precipitation for the same periods as those for collecting, Barlow (1960) was able to determine that *Polydesmus denticulatus* is active over a wide range of temperature (9–20 °C) whereas the two activity periods of *Cylindroiulus frisius* (= *latestriatus*) correspond much more closely to temperature and rainfall (Fig. 10.6). In the three species examined in Barlow's study, activity could be related most clearly to temperature rather than precipitation.

10.7 Responses to human activity

10.7.1 Pollution and its effects on millipedes

There have been few studies that deal with the effects of pollution on millipedes. Even general reports that record effects on terrestrial invertebrates rarely give them more than a passing comment. The majority of studies have concentrated on the passage of pollutants through food

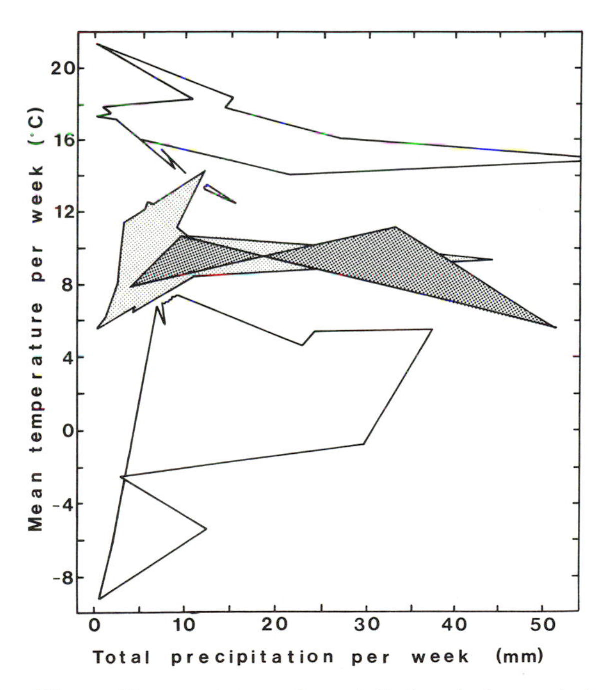

Fig. 10.6 Conditions of temperature and precipitation during periods of maximum activity (spring maximum—light stipple; autumn maximum—dark stipple) and the remainder of the year for *Cylindroiulus latestriatus* in sand dunes in Holland. Redrawn from Barlow (1960).

chains, and accumulation in tissues. Most work has been conducted on metals.

Pollution may affect decomposition in terrestrial ecosystems. For example, Coughtrey *et al.* (1979) postulated that the considerable accumulation of partially decomposed leaf litter that occurs in woodlands polluted by aerial fallout of metals from a zinc, cadmium, and lead smelting works near Avonmouth, England, was due to a lack of detrivorous invertebrates. One of these woodlands lacks julid and glomerid millipedes and has far fewer polydesmids than similar unpolluted sites (Hopkin *et al.* 1985).

Millipedes have been shown to accumulate mercury (Siegel *et al.* 1975), copper and zinc (Beyer *et al.* 1985, 1990; Carter 1983; Hopkin *et al.* 1985; Paoletti *et al.* 1988; Ponomarenko *et al.* 1974), magnesium and strontium (Ponomarenko *et al.* 1974), and cadmium (Hopkin *et al.* 1985). In experiments performed by Hopkin *et al.* (1985) on lead assimilation by *Glomeris marginata*, the millipedes did not accumulate the metal in spite of being fed on a lead-rich diet (Fig. 10.7). This finding is difficult to understand in the

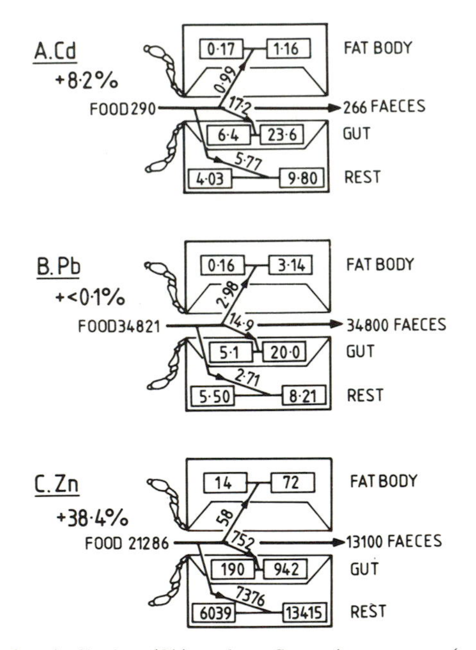

Fig. 10.7 Estimated assimilation (%) and net fluxes in amounts (ng) of metals through *Glomeris marginata*, and between the 'gut', 'fat body', and 'rest' tissue fractions of millipedes from uncontaminated sites fed for 8 weeks on field maple leaf litter (*Acer campestre*) collected from Haw Wood, a site contaminated with cadmium, lead, and zinc. The value given in the left-hand box of each tissue fraction represents the amount of metal estimated to have been present at the start of the experiment. The value in the right-hand box represents the amount present at the end (N = 12). Redrawn from Hopkin *et al.* (1985) by permission of Artis Bibliotheek, Amsterdam.

light of the report of Kohler and Alberti (1991), who found high levels of lead (> 2500 μg g^{-1}) in the intestine of *Glomeris marginata* from a lead-contaminated mine site. Clearly, more research is required to clarify the forms of lead which are available to millipedes in their food.

Loss of metals in the exuvia was considered an important excretory route by Zhulidov and Sizova (1985) despite the fact that millipedes invariably eat their moulted exoskeleton. Zhulidov and Sizova (1985) also reported that consumption of polluted leaf litter in *Sarmatiulus kessleri* was lower, and excretion higher, in comparison with millipedes fed on uncontaminated litter. Consumption was also found to be lower by Read and Martin (1990) when young *Tachypodoiulus niger* and *Glomeris marginata* were fed on leaf litter polluted with zinc, cadmium, and lead. Mortality was also greater in animals fed on polluted litter (Fig. 10.8).

Lower rates of consumption lead to lower growth and result ultimately in decreased populations unless some degree of adaptation occurs. No evidence for the evolution of tolerance to pollutants has been found in millipedes. However, this topic has not been examined experimentally in any detail.

Fluoride accumulation by invertebrates in sites adjacent to a major aerial source of this element was studied by Buse (1986). Millipedes were found to accumulate up to 25 times more fluoride in comparison to the

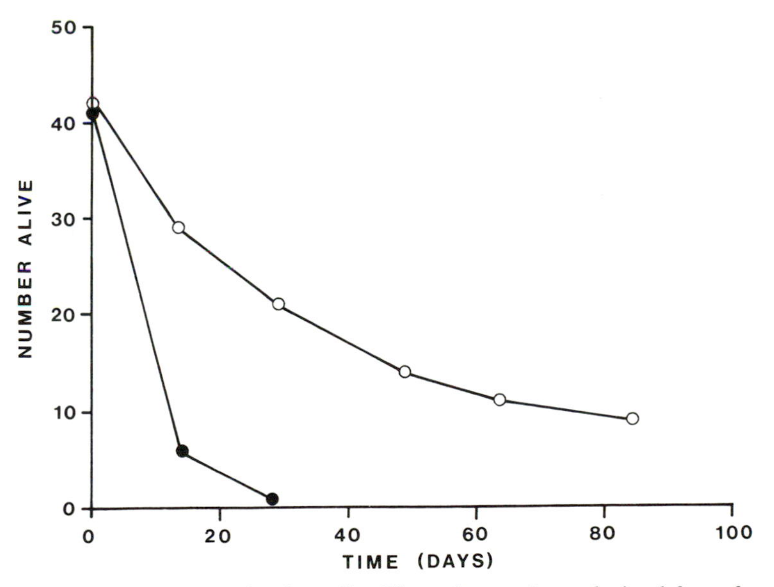

Fig. 10.8 Mortality curves for juvenile *Glomeris marginata* derived from females collected from an uncontaminated site, fed on uncontaminated leaf litter (○), and leaves from Haw Wood (●), a site contaminated with cadmium, lead, and zinc. Redrawn from Read and Martin (1990) by kind permission of E. J. Brill, Leiden.

same species from control sites. At the contaminated site, millipedes contained higher concentrations of fluoride than any other invertebrate studied. Fluoride is a calcium analogue and was almost certainly following the same biochemical pathway in the millipedes. It was probably stored in the calcified cuticle.

Rantala (1990) studied the accumulation of radioactive caesium in *Cylindroiulus britannicus* (Julida) following the Chernobyl disaster of 1986. She found levels of 7360 to 13420 Bq kg^{-1} in animals reared in a radioactive medium. Levels were higher in younger millipedes. Krivoluckij *et al.* (1972) recorded a decrease in numbers of myriapods in experimental forest plots contaminated with Sr^{90} and Krivolutskii and Filippova (1979) showed that that species from arid areas were more resistant to gamma radiation.

10.7.2 Fires

The effect of fires initiated by Man on millipede populations in a cork oak forest has been studied by Saulnier and Athias-Binche (1986). The density of the control population of myriapods was 230 m^{-2}, comparable to other European forests. After burning, the abundance of myriapods decreased by a factor of five, and the biomass by ten times. One year after the fire, the community was still altered in comparison to before the fire. In Australia, populations of the introduced species *Ommatoiulus moreleti* took three years to recover completely from the effects of fire (Baker 1985*c*).

10.8 Factors that determine where millipedes are found

10.8.1 Introduction

Many factors influence where particular species of millipedes are found. Food preferences, resistance to desiccation or waterlogging, temperature preferences, etc. are all important both on a broad scale (i.e. is the species found in deserts or marshland?), or on the smaller scale (is it found in leaf litter or dead wood?).

Barlow (1957) and Haacker (1968) examined the ecology of millipedes in the field, then backed up their observations with laboratory experiments, the results of which, in general, supported the field data. For example, Barlow (1957) discovered that *Schizophyllum (Ommatoiulus) sabulosus*, a wide-ranging species often found in sandy and dry places, is resistant to water loss and is relatively unrestricted by temperature. In contrast, *Cylindroiulus punctatus* has poor resistance and occurs only in a restricted range of habitats. Barlow concluded that humidity probably exerts the greatest influence on distribution but that in some instances, behavioural adaptations could be as important as physiological ones.

The most detailed work relating distribution of millipedes to their environment has been conducted by Kime and co-workers in Belgium. It was concluded, in a study of soil-dwelling species, that distributions were a function of edaphic and climatic factors (Kime *et al.* 1991). The most important were soil texture, soil water content, temperature, mineral content (especially calcium and magnesium), humidity, and humus type (Kime 1991). Ordination procedures have been used to illustrate the relative importance of these factors to a range of species (Kime and Wauthy 1984; Kime *et al.* 1991).

Kime (1991) has also raised the important point that a species on the edge of its range may act differently (in terms of habitat preference) than one in the centre of its range. He recognized four 'species-groups' of millipedes: First, accidental (i.e. usually found in a different habitat); second, sporadic; third, edaphic and on the edge of its range; and fourth, edaphic and within its range (Kime *et al.* 1991).

10.8.2 Niche partitioning and competition

Within any particular site, a range of millipede species usually occurs. As all are invariably detritivores, it may appear that different species occupy the same niche and are in direct competition with each other. However, close scrutiny has not borne this out (Geoffroy *et al.* 1987; Karamaouna 1987; Simonsen 1983).

O'Neill (1967) studied millipedes from seven genera in the USA and found that the precise habitat differed for each. For example, one species was found in heartwood, one under bark, one under logs but on the ground, etc. A rather more detailed study by Enghoff (1983*a*) looked at species of the genus *Cylindroiulus* on Madeira. Microhabitat separated many species but some were found to co-exist, for example, in leaf litter. In these instances, another difference could always be found, for example the size ranges were different, thus influencing the type of food eaten.

Sometimes, two species appear to co-exist which do not differ in size. Such a pair was studied by Geoffroy (1981*b*). *Cylindroiulus punctatus* and *Allajulus (Cylindroiulus) nitidus* can not be separated by size and weight. Both occur in the forest of Foljuif (France) where *Allajulus nitidus* is twice as abundant as *Cylindroiulus punctatus*. The position of the animals within the forest floor is indicated in Fig. 10.9. *Cylindroiulus punctatus* spends the summer in logs and breeds therein. In contrast, *Allajulus nitidus* is in the soil at this time of the year. In the winter *Cylindroiulus punctatus* moves into the soil and *Allajulus nitidus* moves deeper. Thus, for much of the time, they are separated spacially by their seasonal migrations. In spring and autumn, the two species coexist in the F and H layers, but Geoffroy found that they tended to aggregate in monospecific groups. These two species appear to share the same niche for part of the year. However, the forest is mixed and the litter comprises leaves of oak, hornbeam, and pine. In these circumstances, there is the potential for partitioning among food resources. Table 10.1 summarizes the similarities and differences between these two species.

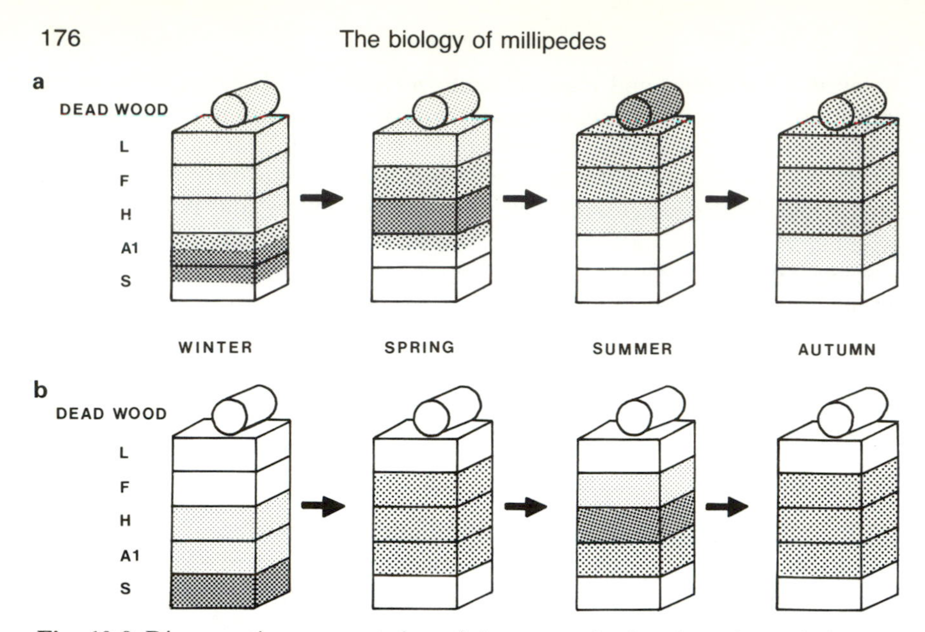

Fig. 10.9 Diagramatic representation of the seasonal migration of populations of (a) *Cylindroiulus punctatus* and (b) *Allajulus (Cylindroiulus) nitidus* in the different compartments of litter, soil and dead wood during the course of a yearly cycle. The density of shading is proportional to the density of the millipedes. Redrawn from Geoffroy (1981*b*) by kind permission of the author and Gauthier-Villars.

Dunger and Steinmetzger (1981) illustrated the change in species that occurs with change in habitat by using kite diagrams (Fig. 10.10). This very visual display emphasizes that small scale changes can be just as dramatic (in terms of the millipede fauna) as large scale ones. The changes observed were related to humidity and pH as well as vegetation type. This study illustrated to some extent the succession of species of millipede that occur during the transition of a site from grassland through scrub to woodland.

Succession of species over time has also been studied by Dunger and Voigtländer (1990). Recolonization of mine sites over 25 years was studied. The millipedes could be divided into four successional groups. *Craspedosoma rawlinsi* was the initial colonizer followed by *Polydesmus inconstans*. The number of species increased gradually whilst some of the early colonizers peaked and declined. On sites more than 20 years old, where woody habitats were becoming established, the species group consisted of up to eight species but was still not comparable to the community found in surrounding 'natural' woods.

A similar type of niche-partitioning (on a larger scale) occurs in some circumstances in relation to altitude. Meyer (1979, 1990) has documented changes in species in the Austrian Alps that are correlated with increased height above sea level. Mauriès (1960) found that species were partitioned

Table 10.1 Comparison of some biological and ecological characteristics of *Allajulus (Cylindroiulus) nitidus* and *Cylindroiulus punctatus* (biomass as mg dry weight). After Geoffroy (1981*b*). Reproduced by kind permission of the author.

Characteristic	*Allajulus nitidus*	*Cylindroiulus punctatus*
Individual biomass at sexual maturity (mg)	♂ 9.7 ♀ 16.2	♂ 9.7 ♀ 16.2
Mean abundance in the superficial soil layers	14 ind^{-2} 115 mg^{-2}	5 ind^{-2} 43 mg^{-2}
Mean abundance in dead wood	0 0	2 ind^{-2} 15 mg^{-2}
Biological cycle	Entirely in leaf litter and soil	Reproduction, egg laying and the first stadia larvae in dead wood
Over wintering	Deep in the soil	In uppermost soil layers
Horizontal distribution in the soil	Aggregated	Aggregated

according to altitude in the French Pyrenees with some species having distinct races at different altitudes (Table 10.2).

10.9 Unusual habitats

10.9.1 Introduction

On going to search for millipedes, the most usual place to try is woodland. Once in the woodland, it then seems most logical to look on the ground, among leaf litter and under logs and stones. However, millipedes are not confined to such places, either in the tropics or in temperate regions. Millipedes can be common in deserts, and various other habitats where one might not expect to find them.

10.9.2 Arboreal millipedes

Perhaps the first place to broaden the search is up in the trees. Many of the active species such as *Tachypodoiulus niger* can be found quite high up trees, using cracks and crevices as daytime resting places. Others such as *Nemasoma varicorne* are always found on wood, often up trees. Enghoff (1982*a*) described the shortened accessory claw on the Madeiran *Cylindroiulus lundbladi* which may be an adaptation for tree climbing.

The most unlikely arboreal millipedes are the giant pill millipedes (Sphaerotheriida). Haacker and Fuchs (1972) described two such species which they found associated with trees in Africa. One occurred on ledges and in holes up to three metres above ground and the other was collected whilst it walked about on the branches. The latter species was seen to climb

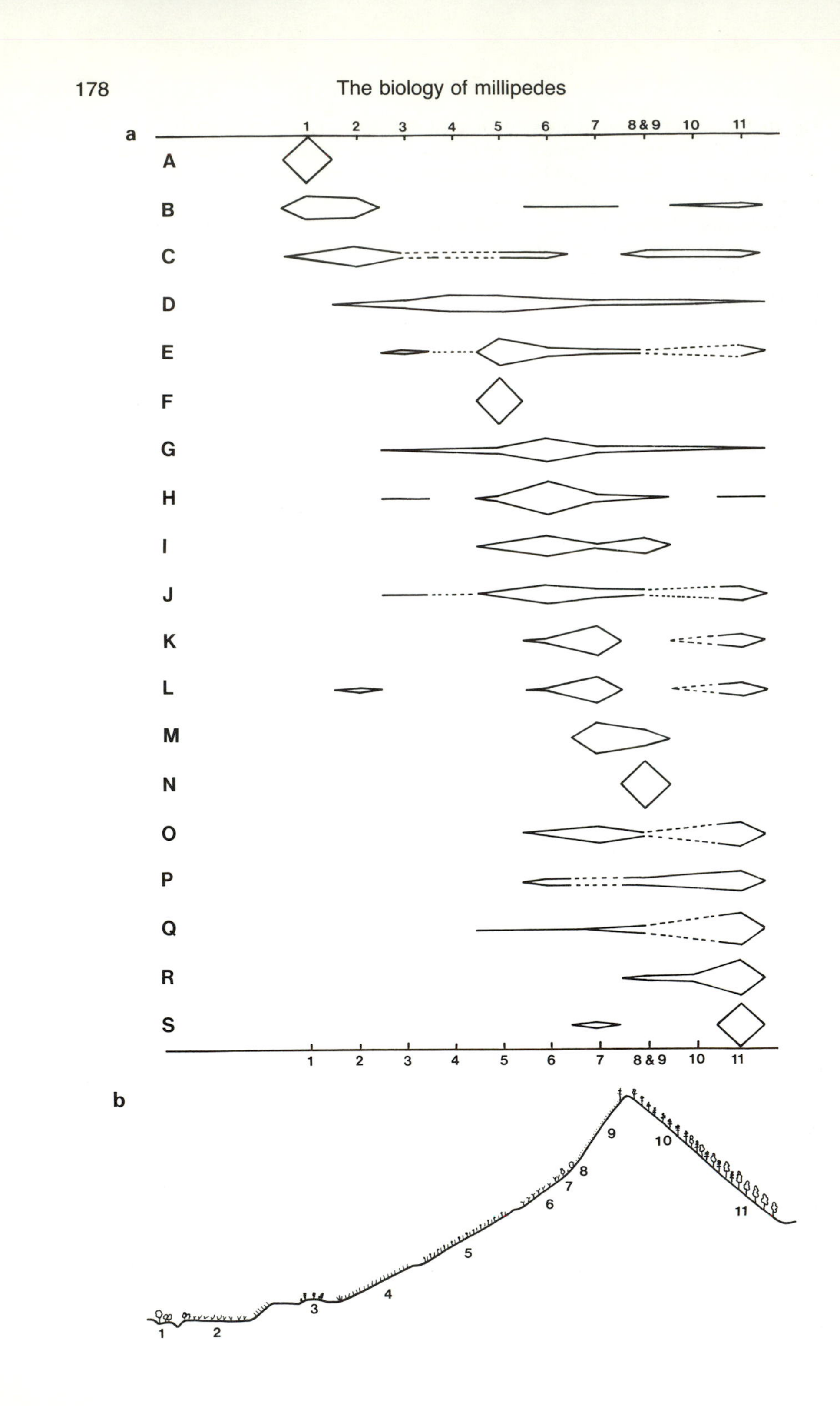
a
1 2 3 4 5 6 7 8 & 9 10 11
A
B
C
D
E
F
G
H
I
J
K
L
M
N
O
P
Q
R
S
1 2 3 4 5 6 7 8 & 9 10 11
b
1
2
3
4
5
6
7
8
9
10
11

Fig. 10.10 (a) Schematic kite diagrams to show the relative abundance and distribution of species of millipede (as determined by captures in pitfall traps) in 11 sites in grassland and woodland (shown in (b) in Thuringia, Germany, between 1971 and 1974. A, *Microchordeuma voigti*; B, *Unciger foetidus*; C, *Polydesmus inconstans*; D, *Cylindroiulus londinensis*; E, *Heteroporatia bosniense*; F, *Polydesmus testaceus*; G, *Glomeris marginata*; H, *Glomeris hexasticha*; I, *Ommatoiulus sabulosus*; J, *Chordeuma silvestre*; K, *Julus scandinavius*; L, *Chromatoiulus unilineatus*; M, *Leptoiulus belgicus*; N, *Blaniulus guttulatus*; O, *Tachypodoiulus niger*; P, *Polydesmus denticulatus*; Q, *Orthochordeuma germanicum*; R, *Polydesmus angustus*; S, *Craspedosoma alemannicum*. Site 1, stream edge; Site 2, water meadow; Site 3, motorway verge; Site 4, lower arable land; Site 5, upper arable land; Site 6, semi-dry grassland; Site 7, scrub zone; Site 8, lower dry grassland; Site 9, upper dry grassland; Site 10, pine forest; Site 11, beech woodland. Redrawn from Dunger and Steinmetzger (1981) by kind permission of the authors and Gustav Fischer Verlag.

---

vertically up and down and was very reluctant to release its grip on the twigs. The usual reaction of sphaerotherids, to roll into a ball, appeared to be inhibited. Analysis of the faeces revealed that the animals had been feeding on soft bark and leaves.

10.9.3 Associations with other animals

Millipedes are often found in close association with other animals. Enghoff (1982*c*) described a collection of millipedes found in a nest of a wood rat. However, one of the most interesting habitats to search is ant nests. The Penicillata (bristly millipedes) seem to be found most frequently in such situations, both in the tropics (e.g. Braxil—Condé 1971) and temperate regions (Washington State, USA—Loomis 1972). The ants are most often *Formica* species, some of which are renowned for their aggression, but this does not appear to concern the millipedes involved.

Ishii and Yamaoka (1982) studied the millipede fauna of arboreal ant nests in a maritime forest in Japan. Of 53 nests examined, 13 contained millipedes with a mean of 2.38 millipedes per nest. Only one species of Polyxenidae was represented.

Another interesting study was carried out by Rettenmeyer (1962) on Barro Colorado Island, Panama. Several species of polydesmid millipedes on the island are true myrmecophiles and run with the army ants (Family Dorylinae). The millipedes live in the nests or bivouacs of the ants and scavenge organic material. They do not appear to eat the faecal pellets but may help to clean the nest and stop mould building up. The ants do not seem to derive any secretion from the millipedes and do not harm them. When the ant colony moves on, as it does every three weeks or so, the millipedes run in the centre of the ant column and may even be carried by the workers. The millipedes can be quite abundant. Emigration from one nest was observed for one hour and 170 millipedes were seen in this time (Chamberlin 1923).

Table 10.2 The distribution of some millipedes in the Massif du Neovieille as a function of altitude. After Mauriès (1960). Reproduced by kind permission of the author.

Site	Beech	Fir	Rhododendron	Lower Cirque of Estarragne	Upper Cirque of Estarragne	Summit of Estarragne
Altitude (m)	1300	1700	1900	2200	2400	2800
Leucoioulus spinosus	√					
Cylindroiulus sagittarius	√					
Blaniulus dollfusi	√	√				
Haplopodoiulus spathifer	√	√				
Cylindroiulus finitimus	√	√				
Chordeuma muticum	√	√				
Hypnosoma exornatum	√	√				
Glomeris marginata	√	√				
Leptoiulus juvenilis	√	√	√			
Glomeris intermedia trisulcata	√	√	√			
Hirudisoma pyrenaeum	√	√	√			
Marquetia lunatum	√	√	√			
Loboglomeris pyrenaicum	√	√	√	√		
Tachypodoiulus niger	√	√	√	√	√	√
Pyreneosoma ribauti					√	
Pyreneosoma digitatum					√	
Ceratosphys guttata					√	

10.9.4 Underwater

An unexpected habitat for millipedes is water. Being adapted to life on land, albeit usually in damp places, it would not be expected that millipedes could survive long periods of submersion. A few studies have shown that this is not necessarily the case. Causey (1943) successfully kept adults of *Oxidus gracilis* underwater for five to seven days but Zulka (1990), studying river flood plain communities, kept *Polydesmus denticulatus* alive for

75 days underwater. This species would appear to survive winter flooding by submersion. A similar finding was made by Adis (1986) where the sub-adults of a tropical species were flooded by the Amazon to a depth of 15 m each year. Adis found that adaptation of 24 hours was needed for the animals to get under the loose bark and acclimatize. During the inundation period, the millipedes emerge to feed on algae (see Section 5.2 for further details).

10.9.5 Caves

Caves are a specialized habitat and when damp, often yield millipedes, especially if there is some type of organic matter present. The usual cavernicolous adaptations can be seen in troglodite millipedes. These include loss of pigmentation, elongation of appendages (legs and antennae), loss of eyes, and a weakening of the cuticle (Chamberlin 1938).

Cave millipedes have been studied extensively by Shear in the USA (Shear 1969, 1984) and Geoffroy in France (Geoffroy 1984). Shear (1984) considered that many cave species in the USA are relics of old groups that sought a better microhabitat during the last Ice Age. Due to the isolation of such habitats, a high degree of endemism is seen in cave millipedes.

The millipedes of waterlogged caves may have modified mouthparts. These have reduced biting parts and enlarged pectinate lamellae which may act as a filter. Such modications have been seen in different species from a variety of different locations (Enghoff 1985).

10.9.6 Deserts

In contrast to the moist habitats so far discussed, millipedes can also be found in desert areas. Crawford and co-workers have made extensive studies of the adaptations and physiology of desert millipedes, especially *Orthoporus ornatus*. As might be expected, adaptation to an arid climate includes a low rate of water loss due to a relatively impermeable cuticle, waxy epicuticle, ability to take up water from unsaturated air, protection of eggs in pellets to prevent water loss, and use of cracks, burrows, stones, etc. to ameliorate the wide fluctuations in temperature that occur in deserts.

The most active periods coincide closely with periods of rain and the animals are able to pass into a dormant state many times during their life. Behavioural adaptations may occur also. For example, individuals may thermoregulate by alternate basking and retreating into cool shelters (Crawford 1979; Crawford *et al.* 1987). Only a few species of millipedes occur in arid regions, the spirostreptids being probably the largest and most obvious. In fact their large size may be an advantage as heat gain and loss is less rapid, as is water loss.

10.9.7 Synanthropic sites

So far, we have seen that millipedes are found in a wide range of natural habitats, but they can also be abundant in unnatural ones too. Heated glasshouses are often promising localities for unusual species and occasionally the animals can reach pest proportions (Edwards and Gunn 1961). Many urban situations such as rubbish dumps, piles of rubble, etc. can be quite promising, a sentiment with which many members of the British Myriapod Group will agree! One of us (S. P. H.) found the first confirmed British record for *Nopoiulus kochii* under rubbish on wasteland adjacent to a motorway in Manchester. However, this site has now disappeared under the new Computer Science building of Manchester University, illustrating that many synanthropic sites are as ephemeral as they are interesting.

Concluding remarks

Millipedes are fascinating and, dare we say it, endearing creatures. Much progress has been made since Owen (1742) classified them as snakes! Nevertheless, many exciting discoveries are waiting to be made both by biologists using modern techniques such as electron microscopy and molecular biology, and by fieldworkers using apparatus no more complicated than a set of pitfall traps. A recent example is that of Pass (1991) who has provided interesting information on possible evolutionary relationships between Onychophora, Myriapoda and Hexapoda by studying their antennal circulatory organs.

As a final point, we would like to make a plea for the study of millipedes for their own sake. One look at scanning electron micrographs of *Polyxenus lagurus* (Fig. 6.5) will convince sceptics that this is one area where science meets art!

References

Achar, K. P. (1986). Analysis of male meiosis in seven species of Indian pill millipedes (Diplopoda: Myriapoda). *Caryologia,* **39,** 89–101.

Achar, K. P. (1987). Chromosomal evolution in Diplopoda (Myriapoda: Arthropoda). *Caryologia,* **40,** 145–55.

Adamson, M. L. (1985). Rhigonematida (Nematoda) of *Rhinocricus bernhardinensis* (Rhinocricidae; Spirobolida; Diplopoda) with comments on *r*- and *K*-selection in nematode parasites of diplopods. *Revue Suisse de Zoologie,* **92,** 871–96.

Adamson, M. L. (1987*a*). Rhigonematid (Rhigonematida: Nematoda) parasites of *Scaphiostreptus seychellarum* (Spirostreptida: Diplopoda) in the Seychelles with comments on ovejector structure in *Rhigonema* Cobb, 1898. *Canadian Journal of Zoology,* **65,** 1889–946.

Adamson, M. L. (1987*b*). Nematode parasites of *Orthoporus americanus* (Diplopoda; Spirobolida) from Paraguay. *Canadian Journal of Zoology,* **65,** 3011–19.

Adamson, M. L. and Van Waerebeke, D. (1985). The Rhigonematida (Nematoda) of Diplopoda: reclassification and its cladistic representation. *Annales de Parasitologie Humaine et Comparee,* **60,** 685–702.

Adis, J. (1986). An 'aquatic' millipede from a Central Amazonian inundation forest. *Oecologia,* **68,** 347–9.

Akre, R. D. and Rettenmeyer, C. W. (1968). Trail following by guests of army ants (Hymenoptera: Formicidae: Ecitonini). *Journal of the Kansas Entomological Society,* **41,** 165–74.

Almond, J. E. (1985*a*). The Silurian–Devonian fossil record of the Myriapoda. *Philosophical Transactions of the Royal Society of London, Series B,* **309,** 227–37.

Almond, J. E. (1985*b*). Les Arthropleurides due Stephanien de Montceau-Les-Mines, France. *Bulletin Société d'Histoire Naturelle d'Autun,* **115,** 59–60.

Altner, H. and Prillinger, L. (1980). Ultrastructure of invertebrate chemo-, thermo-, and hygroreceptors and its functional significance. *International Review of Cytology,* **67,** 69–139.

Anderson, J. M. (1988). Invertebrate-mediated transport processes in soils. *Agriculture, Ecosystems and Environment,* **24,** 5–19.

Anderson, J. M. and Bignell, D. E. (1980). Bacteria in the food, gut contents and faeces of the litter-feeding millipede *Glomeris marginata* (Villers). *Soil Biology and Biochemistry,* **12,** 251–4.

Anderson, J. M. and Bignell, D. E. (1982). Assimilation of ^{14}C-labelled leaf fibre by the millipede *Glomeris marginata* (Diplopoda: Glomeridae). *Pedobiologia,* **23,** 120–5.

Anderson, J. M. and Ineson, P. (1983). Interactions between soil arthropods and microbial populations in carbon, nitrogen and mineral nutrient fluxes from decomposing leaf litter. In *Nitrogen as an ecological factor,* (ed. J. Lee and S. McNeill), pp. 413–32. Blackwell Scientific, Oxford.

Anderson, J. M. and Ineson, P. (1984). Interactions between microorganisms and soil invertebrates in nutrient flux pathways of forest ecosystems. In *Invertebrate–microbial interactions,* British Mycological Society Symposium No. 6, (ed. J. M. Anderson, A. D. M. Rayner, and D. W. H. Walton), pp. 59–88. Cambridge University Press.

Anderson, J. M. and Leonard, M. A. (1988). Tree root and macrofauna effects on nitrification and mineral nitrogen losses from deciduous leaf litter. *Revue d'Ecologie et de Biologie du Sol.,* **25,** 373–84.

Anderson, J. M., Ineson, P., and Huish, S. A. (1983). Nitrogen and cation mobilization by soil fauna feeding on leaf litter and soil organic matter from deciduous woodlands. *Soil Biology and Biochemistry,* **15,** 463–7.

Anderson, J. M., Leonard, M. A., Ineson, P., and Huish, S. (1985). Faunal biomass: a key component of a general model of nitrogen mineralization. *Soil Biology and Biochemistry,* **17,** 735–7.

Andre, M. (1943). Acariens recontrés sur des Myriapodes. *Bulletin du Muséum National d'Histoire Naturelle, Série II,* **15,** 181–5.

Ansenne, A., Compere, P., and Goffinet, G. (1990). Ultrastructural organization and chemical composition of the mineralized cuticle of *Glomeris marginata* (Myriapoda, Diplopoda). In *Proceedings of the 7th International Congress of Myriapodology,* (ed. A. Minelli), pp. 125–34. E. J. Brill, Leiden.

Aouti, A. (1978). Etude comparée des peuplements de Myriapodes Diplopodes d'une forêt hygrophile et d'une plantation d'Hévéa en Basse Côte d'Ivoire. *Annales de l'Université d'Abidjan, Série E,* **11,** 7–32.

Aouti, A. (1980). Monstruosités observées chez des larves pupoides de *Pachybolus laminatus* Cook, Myriapodes Diplopodes. *Annales de l'Université d'Abidjan, Série C,* **16,** 131–6.

Appel, A. G. (1988). Water relations and desiccation tolerance of migrating garden millipedes (Diplopoda: Paradoxosomatidae). *Environmental Entomology,* **17,** 463–6.

Attems, C. (1926). Myriapoda. *Handbuch der Zoologie (Berlin),* **4,** 1–402.

Ax, P. (1984). *The phylogenetic system. The systematization of organisms on the basis of their phylogenesis.* Wiley, New York (in German).

Ax, P. (1987). *The phylogenetic system: the systematization of organisms on the basis of their phylogenies.* Wiley, New York (in German).

Baccetti, B. and Dallai, R. (1978). The evolution of myriapod spermatozoa. *Abhandlungen und Verhandlungen des Naturwissenschaftlichen Vereins in Hamburg,* **21/22,** 203–17.

Baccetti, B., Dallai, R., Bernini, F., and Mazzini, M. (1974). The spermatozoon of Arthropoda XXIV. Sperm metamorphosis in the diplopod *Polyxenus. Journal of Morphology,* **143,** 187–246.

Baccetti, B., Burrini, A. G., Dallai, R., Pallini, V., Camatini, M., Franchi, E., and Paoletti, L. (1977). The 'delayed flagellum' of millipede sperm is a reacted acrosome (The spermatozoon of Arthropoda XVIII). *Journal of Submicroscopic Cytology,* **9,** 187–219.

Baccetti, B., Burrini, A. G., Dallai, R., and Pallini, V. (1979). Recent work in myriapod spermatology (The spermatozoon of Arthropoda XXXI). In *Myriapod biology,* (ed. M. Camatini), pp. 97–104. Academic Press, London.

Bailey, P. T. (1989). The millipede parasitoid *Pelidnoptera nigripennis* (F.) (Diptera: Sciomyzidae) for the biological control of the millipede *Ommatoiulus moreleti* (Lucas) (Diplopoda; Julida; Julidae) in Australia. *Bulletin of Entomological Research,* **79,** 381–91.

Bailey, P. T. and de Mendonca, T. R. (1990). The distribution of the millipede *Ommatoiulus moreleti* (Diplopoda; Julida; Julidae) in relation to other *Ommatoiulus* species on the south-western Iberian Peninsula. *Journal of Zoology,* **221,** 99–111.

Baker, A. N. (1974). Some aspects of the economic importance of millipedes. *Symposia of the Zoological Society of London,* **32,** 621–8.

Baker, G. H. (1978*a*). The post-embryonic development and life history of the millipede *Ommatoiulus moreleti* (Diplopoda: Iulidae) introduced in south-eastern Australia. *Journal of Zoology,* **186,** 209–28.

Baker, G. H. (1978*b*). The population dynamics of the millipede *Ommatoiulus moreleti* (Diplopoda: Iulidae). *Journal of Zoology,* **186,** 229–42.

Baker, G. H. (1978*c*). The distribution and dispersal of the introduced millipede *Ommatoiulus moreleti* (Diplopoda: Iulidae) in Australia. *Journal of Zoology,* **185,** 1–11.

Baker, G. H. (1979*a*). Eruptions of the introduced millipede, *Ommatoiulus moreleti* (Diplopoda: Iulidae), in Australia, with notes on the native *Australiosoma castaneum* (Diplopoda, Paradoxosomatidae). *South Australian Naturalist,* **53,** 36–41.

Baker, G. H. (1979*b*). The activity patterns of *Ommatoiulus moreleti* (Diplopoda: Iulidae) in South Australia. *Journal of Zoology,* **188,** 173–83.

Baker, G. H. (1980). The water and temperature relationships of *Ommatoiulus moreleti* (Diplopoda: Iulidae). *Journal of Zoology,* **190,** 97–108.

Baker, G. H. (1984). Distribution, morphology and life history of the millipede *Ommatoiulus moreleti* (Diplopoda: Iulidae) in Portugal and comparisons with Australian populations. *Australian Journal of Zoology,* **32,** 811–22.

Baker, G. H. (1985*a*). Parasites of the millipede *Ommatoiulus moreleti* (Lucas) (Diplopoda: Iulidae) in Portugal, and their potential as biological control agents in Australia. *Australian Journal of Zoology,* **33,** 23–32.

Baker, G. H. (1985*b*). Predators of *Ommatoiulus moreleti* (Lucas) (Diplopoda: Iulidae) in Portugal and Australia. *Journal of the Australian Entomological Society,* **24,** 247–52.

Baker, G. H. (1985*c*). The distribution and abundance of the Portuguese millipede *Ommatoiulus moreleti* (Diplopoda: Iulidae) in Australia. *Australian Journal of Ecology,* **10,** 249–59.

Balazuc, J. and Schubart, O. (1962). La teratologie des myriapodes. *Annales Biologiques,* **66,** 145–74.

Baleux, B. and Vivares, C. P. (1974). Etude preliminaire de la flore bactérienne intestinale de *Schizophyllum sabulosum* var. *rubripes* Lat. (Myriapoda—Diplopoda). *Bulletin de la Société Zoologique de France,* **99,** 771–9.

Bandyopadhyay, S. and Mukhopadhyaya, M. C. (1988). Distribution of 2 species of polydesmid millipedes *Orthomorpha coarctata* and *Streptogonopus phipsoni* in the grasslands and Taylor's Power Law. *Pedobiologia,* **32,** 7–10.

Banerjee, B. (1967*a*). Diurnal and seasonal variations in the activity of the millipedes *Cylindroiulus punctatus* (Leach), *Tachypodoiulus niger* (Leach) and *Polydesmus angustus* Latzel. *Oikos,* **18,** 141–4.

Banerjee, B. (1967*b*). Seasonal changes in the distribution of the millipede *Cylindroiulus punctatus* (Leach) in decaying logs and soil. *Journal of Animal Ecology,* **36,** 171–7.

Banerjee, B. (1973). The breeding biology of *Polydesmus angustus* Latzel (Diplopoda, Polydesmidae). *Norsk Entomologisk Tidsskrift,* **20,** 291–4.

Bano, K. and Krishnamoorthy, R. V. (1979). Circadian rhythms in the sociability and locomotor activities of the millipede, *Jonespeltis splendidus. Behavioral and Neural Biology,* **25,** 573–82.

Barber, A. D. and Fairhurst, C. P. (1974). A habitat and distribution recording scheme for Myriapoda and other invertebrates. *Symposia of the Zoological Society of London,* **32,** 611–19.

Barlow, C. A. (1957). A factorial analysis of distribution in three species of diplopods. *Tijdschrift voor Entomologie,* **100,** 349–426.

Barlow, C. A. (1960). Distributional and seasonal activity in three species of diplopods. *Archives Néerlandaises de Zoologie,* **13,** 108–33.

Barnwell, F. H. (1965). An angle sense in the orientation of a millipede. *Biological Bulletin,* **128,** 33–50.

Beck, L. and Friebe, B. (1981). Verwertung von Kohlenhydraten bei *Oniscus asellus* (Isopoda) und *Polydesmus angustus* (Diplopoda). *Pedobiologia,* **21,** 19–29.

Bedini, C. (1970). The fine structure of the eye in *Glomeris* (Diplopoda). *Monitore Zoologico Italiano* (New Series), **4,** 201–19.

Bedini, C. and Mirolli, M. (1967). The fine structure of the temporal organs of pill millipede *Glomeris romana* Verh. *Monitore Zoologico Italiano* (New Series), **1,** 41–63.

Bellairs, V., Bellairs, R., and Goel, S. (1983). Studies on an Indian polydesmoid millipede *Streptogonopus phipsoni.* Life cycle and swarming behaviour of the larvae. *Journal of Zoology,* **199,** 31–50.

Bennett, D. S. (1971). Nitrogen excretion in the diplopod *Cylindroiulus londinensis. Comparative Biochemistry and Physiology,* **39A,** 611–24.

Bercovitz, K. and Warburg, M. R. (1985). Developmental patterns in two populations of the millipede *Archispirostreptus syriacus* (De Saussure) in Israel (Diplopoda). *Bijdragen tot de Dierkunde,* **55,** 37–46.

Bercovitz, K. and Warburg, M. R. (1988). Factors affecting egg-laying and clutch size of *Archispirostreptus tumuliporus judaicus* (Attems) (Myriapoda), Diplopoda in Israel. *Soil Biology and Biochemistry,* **20,** 869–74.

Bernardi, M. de, Mellerio, G., Vidari, G., Vita-Finzi, P., Demange, J. M., and Pavan, M. (1982). Quinones in the defensive secretions of African millipedes. *Naturwissenschaften,* **69,** 601–2.

Berns, M. W. (1968). The development of the copulatory organs (gonopods) of a Spirobolid millipede. *Journal of Morphology,* **126,** 447–62.

Bertrand, M., Janatiidrissi, A., and Lumaret, J. P. (1987). Etude experimentale des facteurs de variation de la consommation de la litière de *Quercus ilex* L. et *Q. pubescens* Willd. par *Glomeris marginata* (V.) (Diplopoda, Glomeridae). *Revue d'Ecologie et de Biologie du Sol,* **24,** 359–68.

Bessiere, C. (1948). La spermatogenese de quelques Myriapodes Diplopodes. *Archives de Zoologie Expérimentale et Générale,* **85,** 149–239.

Beyer, W. N., Pattee, O. H., Sileo, L., Hoffman, D. J., and Mulhern, B. M. (1985). Metal contamination in wildlife living near two zinc smelters. *Environmental Pollution* (Series A), **38,** 63–86.

Beyer, W. N., Miller, G., and Simmers, J. W. (1990). Trace elements in soil and soil biota in confined disposal facilities for dredged material. *Environmental Pollution,* **65,** 19–32.

Bhakat, S. N. (1987). Ecology of an Indian grassland millipede *Streptogonopus phipsoni* (Diplopoda, Polydesmoidea). *Journal of Zoology,* **212,** 419–28.

Bhakat, S. (1989*a*). Ecology of *Chondromorpha kelaarti* (Diplopoda, Polydesmida). *Journal of Zoology,* **219,** 209–19.

Bhakat, S. (1989*b*). The population ecology of *Orthomorpha coarctata* (Diplopoda: Polydesmidae). *Pedobiologia,* **33,** 49–59.

Bhakat, S., Bhakat, A., and Mukhopadhyaya, M. C. (1989). The reproductive biology and post-embryonic development of *Streptogonopus phipsoni* (Diplopoda: Polydesmoidea). *Pedobiologia,* **33,** 37–47.

Biernaux, J. (1966). Incidence économique des Iules en culture betteravière. *Mededelingen van de Rijksfaculteit Landbouwwetenschappen te Gent,* **31,** 717–29.

Biernaux, J. and Baurant, R. (1964). Observations sur l'hibernation de *Archiboreoiulus pallidus* Br.-Bk. (Myriapode—Diplopode—Iulidae). *Bulletin de l'Institute Agronomique et des Stations de Recherches de Gembloux,* **32,** 290–8.

Biernaux, J., Vincinaux, C., and Seutin, E. (1973). Recherches recentes sur l'efficacite des insecticides contre les 'Iules de la betteraue'. *Mededelingen van de Rijksfaculteit Landbouwwetenschappen te Gent,* **38,** 1187–203.

Bignell. D. E. (1984*a*). Direct potentiometric determination of redox potentials of the gut contents in the termites *Zootermopsis nevadensis* and *Cubitermes severus* and three other arthropods. *Journal of Insect Physiology,* **30,** 169–74.

Bignell, D. E. (1984*b*). The arthropod gut as an environment for microorganisms. In *Invertebrate—Microbial Interactions,* British Mycological Society Symposium No. 6, (ed. J. M. Anderson, A. D. M. Rayner, and D. W. H. Walton), pp. 205–27. Cambridge University Press.

Bignell, D. E. (1989). Relative assimilations of ^{14}C-labelled microbial tissues and ^{14}C-labelled plant fibre ingested with leaf litter by millipede *Glomeris marginata* under experimental conditions. *Soil Biology and Biochemistry,* **21,** 819–28.

Binns, E. S. (1982). Phoresy as migration—some functional aspects of phoresy in mites. *Biological Reviews (and Biological Proceedings) of the Cambridge Philosophical Society,* **57,** 571–620.

Blower, J. G. (1969). Age-structures of millipede populations in relation to activity and dispersion. *Systematics Association Publications,* **8,** 209–16.

Blower, J. G. (1970*a*). The millipedes of a Cheshire wood. *Journal of Zoology,* **160,** 455–96.

Blower, J. G. (1970*b*). Notes on the life histories of some British Iulidae. *Bulletin du Muséeum National d'Histoire Naturelle, Série II,* **41,** 19–23.

Blower, J. G. (1974*a*) (ed.). Myriapoda. *Symposia of the Zoological Society of London,* **32,** 712.

Blower, J. G. (1974*b*). Food consumption and growth in a laboratory population of *Ophyiulus pilosus* (Newport). *Symposia of the Zoological Society of London,* **32,** 527–51.

Blower, J. G. (1978). Anamorphosis in the Nematophora. *Abhandlungen und Verhandlungen des Naturwissenschaftlichen Vereins in Hamburg,* **21/22,** 97–103.

Blower, J. G. (1985). *Millipedes.* Linnean Society Synopses of the British Fauna (New Series), Number 35. E. J. Brill/Dr W. Backhuys, London.

Blower, J. G. and Fairhurst, C. P. (1968). Notes on the life-history and ecology of *Tachypodoiulus niger* (Diplopoda, Iulidae) in Britain. *Journal of Zoology,* **156,** 257–71.

Blower, J. G. & Gabbutt, P. D. (1964). Studies on the millipedes of a Devon oak wood. *Proceedings of the Zoological Society of London,* **143,** 143–76.

Blower, J. G. and Miller, P. F. (1974). The life cycle and ecology of *Ophiulus pilosus* (Newport) in Britain. *Symposia of the Zoological Society of London,* **32,** 503–25.

Blower, J. G. and Miller, P. F. (1977). The life history of the julid millipede *Cylindroiulus nitidus* in a Derbyshire wood. *Journal of Zoology,* **183,** 339–51.

Blum, M. S. and Woodring, J. P. (1962). Secretion of benzaldehyde and hydrogen cyanide by the millipede *Pachydesmus crassicutis* (Wood). *Science,* **138,** 512–13.

Bocock, K. L. (1983). The digestion of food by *Glomeris*. In *Soil organisms*, (ed. J. Doeksen and J. Van de Drift), pp. 85–91. Elsevier–North Holland, Amsterdam.

Boissin, L., Manier, J. F., and Tuzet, O. (1972). Etude ultrastructurale de la spermatogenèse et de la spermiogenèse de *Glomeris marginata* Villers (Myriapode, Diplopode). *Annales des Sciences Naturelles, Zoologie,* **14,** 221–40.

Boudreaux, H. B. (1979). *Arthropod phylogeny with special reference to insects.* Wiley, New York.

Bowen, R. C. (1967). Defence reactions of certain Spirobolid millipedes to larval *Macracanthorhynchus ingens. Journal of Parasitology,* **53,** 1092–5.

Bowie, J. Y. (1985). New species of rhigonematid and thelastomatid nematodes from indigenous New Zealand millipedes. *New Zealand Journal of Zoology,* **12,** 485–503.

Brade-Birks, S. G. (1922). Notes on Myriapoda XXVII. Wandering millipedes. *Annals and Magazine of Natural History, Series 9,* **9,** 208–12.

Brade-Birks, S. G. (1923). Notes on Myriapoda XXIX: a preliminary communication on economic status. *Supplement to the Lancashire and Cheshire Naturalist,* December 1923, 1–8.

Brade-Birks, S. G. (1929). Notes on Myriapoda XXXIII. The economic status of Diplopoda and Chilopoda and their allies. Part I. *Journal of the South-Eastern Agricultural College, Wye, Kent,* **26,** 178–216.

Brade-Birks, S. G. (1930). Notes on Myriapoda XXXIII. The economic status of Diplopoda and Chilopoda and their allies. Part II. *Journal of the South-Eastern Agricultural College, Wye, Kent,* **27,** 103–46.

Brade-Birks, S. G. (1974). Presidential address: retrospect and prospect in myriapodology. *Symposia of the Zoological Society of London,* **32,** 1–12.

Briggs, D. E. G., Plint, A. G., and Pickerill, R. K. (1984). Arthropleura trails from the Westphalian of Eastern Canada. *Palaeontology,* **27,** 843–55.

British Myriapod Group (1988). *Preliminary atlas of the millipedes of the British Isles.* Biological Records Centre, Natural Environment Research Council, Institute of Terrestrial Ecology, Monks Wood Experimental Station, Huntingdon.

Brolemann, H. W. (1935. Myriapodes Diplopodes (Chilognathes I). *Faune de France,* **29,** 1–368. Lechevalier, Paris.

Brookes, C. H. (1974). The life cycle of *Proteroiulus fuscus* (Am Stein) and *Isobates varicornis* (Koch) with notes on the anamorphosis of Blaniulidae. *Symposia of the Zoological Society of London,* **32,** 485–501.

Brooks, C. H. and Willoughby, J. (1978). An investigation of the ecology and life history of the millipede *Blaniulus guttulatus* (Bosc.) in a British woodland. *Abhandlungen und Verhandlungen des Naturwissenschaftlichen Vereins in Hamburg,* **21/22,** 105–14.

Brusca, R. C. and Brusca, G. J. (1990). *Invertebrates.* Sinauer, Sunderland, Massachusetts.

Burtt, E. (1938). Irritant exudation from a millipede. *Nature,* **142,** 796.

Burtt, E. (1947). Exudate from millipedes with particular reference to its injurious effects. *Tropical Diseases Bulletin,* **44,** 7–12.

Buse, A. (1986). Fluoride accumulation in invertebrates near an aluminium reduction plant in Wales. *Environmental Pollution* (Series A), **41,** 199–217.

Calow, P. (1978). *Life cycles.* Chapman & Hall, London.

Camatini, M. (1979). (ed.). *Myriapod biology.* Academic Press, London.

Camatini, M. and Castellani, L. C. (1978). Myofilaments array of some visceral muscle fibers of *Lithobius forficatus* Linnaeus and *Pachyiulus enologus* B. *Abhandlungen und Verhandlungen des Naturwissenschaftlichen Vereins in Hamburg,* **21/22,** 243–55.

Camatini, M. and Franchi, E. (1978). Studies on the fine structure of spermatozoa from the millipede *Pachyiulus enologus* B. *Abhandlungen und Verhandlungen des Naturwissenschaftlichen Vereins in Hamburg,* **21/22,** 231–41.

Candia Carnevali, M. D. and Valvassori, R. (1982). Active supercontraction in rolling-up muscles of *Glomeris marginata* (Myriapoda, Diplopoda). *Journal of Morphology,* **172,** 75–82.

Carey, C. J. and Bull, C. M. (1986). Recognition of mates in the Portugese millipede *Ommatoiulus moreleti. Australian Journal of Zoology,* **34,** 837–42.

Carmignani, M. P. A. and Zaccone, G. (1977). Morphochemical study of the cuticle in the millipede *Pachyiulus flavipes* C. Koch (Diplopoda, Myriapoda). *Cellular and Molecular Biology,* **22,** 163–8.

Carrel, J. E. (1979). Defensive secretion of the pill-millipede *Glomeris marginata*: fluid content of adult individuals. *American Zoologist,* **19,** 886.

Carrel, J. E. (1984). Defensive secretion of the pill millipede *Glomeris marginata* I. Fluid production and storage. *Journal of Chemical Ecology,* **10,** 41–51.

Carrel, J. E. (1990). Chemical defense in the pill millipede *Glomeris marginata.* In *Proceedings of the 7th International Congress of Myriapodology,* (ed. A. Minelli), pp. 157–64. E. J. Brill, Leiden.

Carrel, J. E. and Eisner, T. (1984). Spider sedation induced by defensive chemicals of millipede prey. *Proceedings of the National Academy of Sciences of the United States of America,* **81,** 806–10.

Carrel, J. E., Doom, J. P., and MacCormick, J. P. (1985). Arborine and methaqualone are not sedative in the wolf spider *Lycosa ceratiola* Gertsch and Wallace. *Journal of Arachnology,* **13,** 269–71.

Carter, A. (1983). Cadmium, copper and zinc in soil animals and their food in a red clover system. *Canadian Journal of Zoology,* **61,** 2751–7.

Casnati, G., Nencini, G., Quilico, A., Pavan, M., Ricca, A., and Salvatori, T. (1963). The secretion of the myriapod *Polydesmus collaris collaris* (Koch). *Experentia,* **19,** 409–11.

Caughley, G. (1970). Eruption of ungulate populations with emphasis on Himalayan thar (*Hemitragus jemlahicus*) in New Zealand. *Ecology,* **51,** 53–72.

Causey, N. B. (1943). Studies on the life history and the ecology of the hothouse millipede, *Orthomorpha gracilis* (C. L. Koch 1847). *American Midland Naturalist,* **29,** 670–82.

Causey, N. B. and Tiemann, D. L. (1969). A revision of the bioluminescent millipedes of the genus *Motyxia* (Xystodesmidae, Polydesmida). *Proceedings of the American Philosophical Society,* **113,** 14–33.

Chamberlin, R. V. (1923). On four termitophilous millipedes from British Guiana. *Zoologica,* **3,** 411–21.

Chamberlin, R. V. (1938). Diplopoda from Yucatan. In *Fauna of the caves of yucatan,* (ed. A. S. Pearse), pp. 165–82. Carnegie Institution of Washington Publication No. 491.

Chu, T. L., Szabó, I. M., and Szabó, I. (1987). Nocardioform gut actinomycetes of *Glomeris hexasticha* Brandt (Diplopoda). *Biology and Fertility of Soils,* **3,** 113–16.

Churchfield, J. S. (1979). A note on the diet of the European Water Shrew, *Neomys fodiens bicolor*. *Journal of Zoology,* **188,** 294–6.

Citernesi, U., Neglia, R., Seritti, A., Lepidi, A. A., Filippi, C., Bagnoli, G., Nuti, M. P., and Galluzzi, R. (1977). Nitrogen fixation in the gastro-enteric cavity of soil animals. *Soil Biology and Biochemistry,* **9,** 71–2.

Cloudsley-Thompson, J. L. (1949). The significance of migration in Myriapods. *Annals and Magazine of Natural History, Series 12,* **2,** 947–62.

Cloudsley-Thompson, J. L. (1950). Economics of the 'spotted snake millipede' *Blaniulus guttulatus* Bosc. *Annals and Magazine of Natural History, Series 12,* **3,** 1047–57.

Cloudsley-Thompson, J. L. (1951). On the responses to environmental stimuli, and the sensory physiology of millipedes (Diplopoda). *Proceedings of the Zoological Society of London,* **121,** 253–77.

Cloudsley-Thompson, J. L. (1988). *Evolution and Adaptation of Terrestrial Arthropods.* Springer-Verlag, Berlin.

Condé, B. (1971). Diplopodes Pénicillates des nids brésiliens de *Camponotus rufipes. Revue d'Ecologie et de Biologie du Sol,* **8,** 631–4.

Conner, W. E., Jones, T. H., Eisner, T., and Meinwald, J. (1977). Benzoyl cyanide in the defensive secretion of a polydesmoid millipedes. *Experentia,* **33,** 206–7.

Cook, O. F. (1900). Camphor secreted by an animal (*Polyzonium*). *Science*, **12,** 516–21.

Coolidge, K. R. (1909). Secretion of hydrocyanic acid by *Leptodesmus haydenianus,* Wood. *Canadian Entomologist,* **41,** 104.

Coughtrey, P. J., Jones, C. H., Martin, M. H., and Shales, S. W. (1979). Litter accumulation in woodlands contaminated by Pb, Zn, Cd and Cu. *Oecologia,* **39,** 51–60.

Crane, D. F. and Cowden, R. R. (1968). A cytochemical study of oocyte growth in four species of millipedes. *Zeitschrift für Zellforschung und Mikroskopische Anatomie,* **90,** 414–31.

Crawford, C. S. (1972). Water relations in a desert millipede *Orthoporus ornatus* (Girard) (Spirostreptidae). *Comparative Biochemistry and Physiology,* **42A,** 521–35.

Crawford, C. S. (1976). Feeding-season production in the desert millipede *Orthoporus ornatus* (Girard) (Diplopoda). *Oecologia,* **24,** 265–76.

Crawford, C. S. (1978). Seasonal water balance in *Orthoporus ornatus*, a desert millipede. *Ecology,* **59,** 996–1004.

Crawford, C. S. (1979). Desert millipedes: a rationale for their distribution. In *Myriapod biology,* (ed. M. Camatini), pp. 171–81. Academic Press, London.

Crawford, C. S. (1988). Nutrition and habitat selection in desert detrivores. *Journal of Arid Environments,* **14,** 111–22.

Crawford, C. S. and Matlack, M. C. (1979). Water relations of desert millipede larvae, larva-containing pellets, and surrounding soil. *Pedobiologia,* **19,** 48–55.

Crawford, C. S. and Warburg, M. R. (1982). Water balance and apparent oocyte resorption in desert millipedes. *Journal of Experimental Zoology,* **222,** 215–26.

Crawford, C. S., Minion, G. P., and Bayers, M. D. (1983). Intima morphology, bacterial morphotypes, and effects of annual molt on microflora in the hindgut of the desert millipede *Orthoporus ornatus* (Girard) (Diplopoda: Spirostreptidae). *International Journal of Insect Morphology and Embryology,* **12,** 301–12.

Crawford, C. S., Goldenberg, S., and Warburg, M. R. (1986). Seasonal water balance in *Archispirostreptus syriacus* (Diplopoda: Spirostreptidae) from mesic and xeric Mediterranean environments. *Journal of Arid Environments,* **10,** 127–36.

Crawford, C. S., Bercovitz, K., and Warburg, M. R. (1987). Regional environments, life-history patterns and habitat use of Spirostreptid millipedes in arid regions. *Zoological Journal of the Linnean Society of London,* **89,** 63–88.

Critchley, B. R., Cook, A. G., Critchley, U., Perfect, T. J., Russel-Smith, A., and Yeadow, R. (1979). Effects of bush clearing and soil cultivation on the invertebrate fauna of a forest soil in the humid tropics. *Pedobiologia,* **19,** 425–38.

Cromack, K., Sollins, P., Todd, R. L., Crossley, D. A., Fender, W. M., Fogel, R., and Todd, A. W. (1977). Soil-microorganism–arthropod interactions: Fungi as major calcium and sodium sources. In *The role of arthropods in forest ecosystems,* (ed. W. J. Mattson), pp. 78–84. Springer-Verlag, Berlin.

Curtis, J. (1860). *Farm insects.* Blackie, London.

Dallai, R., Bigliardi, E., and Lane, N. J. (1990). Intercellular junctions in myriapods. *Tissue and Cell,* **22,** 359–69.

Davenport, D., Wooten, D. M., and Cushing, J. E. (1952). The biology of the Sierra luminous millipede *Luminodesmus sequoiae,* Loomis and Davenport. *Biological Bulletin,* **102,** 100–10.

David, J. F. (1982). Variabilité dans l'espace et dans le temps des cycles de vie de deux populations de *Cylindroiulus nitidus* (Verhoeff) (Iulida). *Revue d'Ecologie et de Biologie du Sol.,* **19,** 411–25.

David, J. F. (1984). Le cycle annuel du Diplopode *Microchordeuma gallicum* (Latzel, 1884). *Bulletin de la Société Zoologique de France,* **109,** 61–70.

David, J. F. (1987*a*). Consommation annuelle d'une litière de chêne par une population adulte du Diplopode *Cylindroiulus nitidus. Pedobiologia,* **30,** 299–310.

David, J. F. (1987*b*). Relations entre les peuplements de Diplopodes et les temps d'humus en forêt d'Orleans. *Revue d'Ecologie et de Biologie du Sol,* **24,** 515–25.

David, J. F. (1991). Some questions about the evolution of life-history traits in Diplopoda. *Proceedings of the 8th International Congress of Myriapodology. Veröffentlichungen der Universität Innsbruck.* (In press.)

David, J. F. and Couret, T. (1983). Le développement post-embryonnaire en conditions naturelles de *Polyzonium germanicum* (Brandt) (Diplopoda, Polyzoniida) les cinq premieres stades. *Bulletin du Muséum National d'Histoire Naturelle, Série IV,* **5A,** 585–90.

David, J. F. and Couret, T. (1984). La fin du developpement post-embryonnaire en conditions naturelles de *Polyzonium germanicum* (Brandt) (Diplopoda, Polyzoniida). *Bulletin du Muséum National d'Histoire Naturelle, Série IV,* **6A,** 1067–76.

Demange, J. M. (1972). Contribution à la connaissance du développement post-embryonnaire de *Pachybolus ligulatus* (Voges) (Développement segmentaire, croissance ocellaire, croissance des organes copulateurs, notion de lignées larvaires, zone de croissance). *Biologia Gabonica,* **8,** 127–61.

Demange, J. M. (1974). Réflexions sur le développement de quelques diplopodes. *Symposia of the Zoological Society of London,* **32,** 273–87.

Demange, J. M. (1981). *Les mille-pattes.* Boubée, Paris.

Demange, J. M. (1988). Arthropoda—Myriapoda. In *Reproductive biology of invertebrates, Volume III, Accessory sex glands,* (ed. K. G. Adiyodi and R. G. Adiyodi), pp. 473–85. Wiley, New York.

Demange, J. M. and Gasc, C. (1972). Examen des matériaux rassembles pour une contribution à l'étude du développement post-embryonnaire de *Pachybolus ligulatus* (Voges). *Biologia Gabonica,* **8,** 163–73.

De Mets, R. (1962). Submicroscopic structure of the peritrophic membrane in arthropods. *Nature,* **196,** 77–8.

Dimelow, E. J. (1963). Observations on the feeding of the hedgehog (*Erinaceus europaeus* L.). *Proceedings of the Zoological Society of London,* **141,** 291–309.

Dohle, W. (1964). Die Embryonalentwicklung von *Glomeris marginata* (Villers) im Vergleich zur Entwicklung anderer Diplopoden. *Zoologische Jahrbücher* (*Anatomie*), **81,** 241–310.

Dohle, W. (1965). Über die Stellung der Diplopoden im System. *Zoologischer Anzeiger (Supplement),* **28,** 597–606.

Dohle, W. (1974*a*). The origin and inter-relations of the myriapod groups. *Symposia of the Zoological Society of London,* **32,** 191–8.

Dohle, W. (1974*b*). The segmentation of the germ band of Diplopoda compared with other classes of arthropods. *Symposia of the Zoological Society of London,* **32,** 143–61.

Dohle, W. (1979). Vergleichende Entwicklungsgeschichte des Mesodermis bei Articulaten. *Zeitschrift für Zoologische Systematik und Evolutionsforschung,* **1,** 120–40.

Dohle, W. (1980). Sind die Myriapoden eine monophyletische Gruppe? Eine Diskussion der Verwandtschaftsbeziehungen der Antennaten. *Abhandlungen und Verhandlungen des Naturwissenschaftlichen Vereins in Hamburg,* **23,** 45–104.

Dohle, W. (1988). *Myriapoda and the ancestry of insects.* Manchester Polytechnic.

Dollfus, R. P. (1952). Quelques Oxyuroidea de Myriapodes. *Annales de Parasitologie Humaine et Comparée,* **27,** 143–236.

Dondale, C. D., Redner, J. H., and Semple, R. B. (1972). Diel activity periodicities in meadow arthropods. *Canadian Journal of Zoology,* **50,** 1155–63.

Donisthorpe, H. J. K. (1927). *The guests of British ants: their habits and life-histories.* Routledge, London.

Duffey, S. S. and Blum, M. S. (1977). Phenol and guaiacol: biosynthesis, detoxication and function in a polydesmid millipede *Oxidus gracilis. Insect Biochemistry,* **7,** 57–65.

Duffey, S. S., Underhill, E. W., and Towers, G. H. N. (1974). Intermediates in the biosynthesis of HCN and benzaidehyde by a polydesmid millipede *Harpaphe haydeniana. Comparative Biochemistry and Physiology,* **47B,** 753–66.

Duffey, S. S., Blum, M. S., Fales, H. M., Evans, S. L., Roncadori, W., Tiemann, D. L., and Nakagawa, Y. (1977). Benzol cyanide and mandelonitrile benzoate in the defensive secretions of millipedes. *Journal of Chemical Ecology,* **3,** 101–13.

Duffield, R. M., Blum, M. S., and Brand, J. M. (1974). Guaicol in the defensive secretions of polydesmid millipedes (Myriapoda, Diplopoda, Paradoxosomatidae). *Annals of the Entomological Society of America,* **67,** 821–2.

Dunger, W. and Steinmetzger, K. (1981). Ecological investigations on Diplopoda of a grassland-wood-catena in a limestone area in Thuringia (G.D.R). *Zoologische Jahrbücher (Systematik),* **108,** 519–53.

Dunger, W. and Voigtländer, K. (1990). Succession of Myriapoda in primary colonization of reclaimed land. In *Proceedings of the 7th International Congress of Myriapodology,* (ed. A. Minelli), pp. 219–27. E. J. Brill, Leiden.

Dwarakanath, S. K. (1971). The influence of body size and temperature upon the oxygen consumption in the millipede *Spirostreptus asthenes* (Pocock). *Comparative Biochemistry and Physiology,* **38A,** 351–8.

Dzik, J. (1975). Spirobolid millipedes from the late Cretaceous of the Gobi desert, Mongolia. *Palaeontologia Polonica,* **33,** 17–24.

Dzik, J. (1981). An early Triassic millipede from Siberia and its evolutionary significance. *Neues Jahrbüch für Geologie und Palaeontologie, Monatshefte,* **7,** 395–404.

Dzingov, A., Márialigeti, K., Jáger, K., Contreras, E., Kondics, L., and Szabó, I. M. (1982). Studies on the microflora of millipedes (Diplopoda). I. A comparison of actinomycetes isolated from surface structures of the exoskeleton and digestive tract. *Pedobiologia,* **24,** 1–7.

Edney, E. B. (1951). The evaporation of water from woodlice and the millipede *Glomeris. Journal of Experimental Biology,* **28,** 91–115.

Edney, E. B. (1977). *Water balance in land arthropods.* Springer-Verlag, Heidelberg.

Edwards, C. A. (1974). Some effects of insecticides on myriapod populations. *Symposia of the Zoological Society of London,* **32,** 645–55.

Edwards, C. A. and Gunn, E. (1961). Control of the glasshouse millipede. *Plant Pathology,* **10,** 21–4.

Eisenbeis, G. and Wichard, W. (1987). *Atlas on the biology of soil arthropods.* Springer-Verlag, Berlin.

Eisner, T. (1968). Mongoose and millipedes. *Science,* **160,** 1367.

Eisner, T. (1970). Chemical defense against predation in arthropods. In *Chemical ecology,* (ed. E. Sondheimer and J. B. Simeone), pp. 157–217. Academic Press, New York.

Eisner, T. and Davis, J. A. (1967). Mongoose throwing and smashing millipedes. *Science,* **155,** 577–9.

Eisner, T. and Meinwald, J. (1966). Defensive secretions of arthropods. *Science,* **153,** 1341–50.

Eisner, T., Eisner, H. E., Hurst, J. J., Kafatos, F. C., and Meinwald, J. (1963). Cyanogenic glandular apparatus of a millipede. *Science,* **139,** 1218–20.

Eisner, H. E., Alsop, D. W., and Eisner, T. (1967). Defense mechanisms of arthropods XX. Quantitative assessment of hydrogen cyanide production in two species of millipedes. *Psyche,* **74,** 107–17.

Eisner, T., Zahler, S. A., Carrel, J. E., Brown, D. J., and Lones, G. W. (1970). Absence of antimicrobial substances in the egg capsules of millipedes. *Nature,* **225,** 661.

Eisner, T., Alsop, D., Hicks, K., and Meinwald, J. (1978). Defensive secretions of millipedes. In *Arthropod venoms* (Handbook of Pharmacology No. 48), (ed. S. Bettini), pp. 41–72. Springer-Verlag, Berlin.

El-Hifnawi, E. S. (1973). Topographie und Ultrastruktur der Maxillarnephridien von Diplopoden. *Zeitschrift für Wissenschaftliche Zoologie,* **186,** 118–48.

El-Hifnawi, E. S. and Seifert, G. (1972). Elektronenmikroskopische und experimentelle Untersuchungen uber die Kragendruse von *Polyxenus lagurus* (L.) (Diplopoda, Penicillata). *Zeitschrift für Zellforschung und Mikroskopische Anatomie,* **131,** 255–68.

Elliott, H. J. (1970). The role of millipedes in the decomposition of *Pinus radiata* litter in the Australian Capital territory. *Australian Forest Research,* **4,** 3–10.

Enghoff, H. (1976*a*). Morphological comparison of bisexual and parthenogenetic *Polyxenus lagurus* (Linne, 1758) in Denmark and Southern Sweden, with notes on taxonomy, distribution and ecology. *Entomologiske Meddeleleser,* **44,** 161–82.

Enghoff, H. (1976*b*). Parthenogenesis and bisexuality in the millipede *Nemasoma varicorne* C. L. Koch 1847 (Diplopoda: Blaniulidae). Morphological, ecological and biogeographical aspects. *Videnskabelige Meddelelser fra Dansk Naturhistorisk Forening i Kjobenhavn,* **139,** 21–59.

Enghoff, H. (1976*c*). Competition in connection with geographic parthenogenesis. Theory and examples, including some original observations on *Nemasoma varicorne* C. L. Koch (Diplopoda: Blaniulidae). *Journal of Natural History,* **10,** 475–9.

Enghoff, H. (1978*a*). Parthenogenesis and spanandry in millipedes. *Abhandlungen und Verhandlungen des Naturwissenschaftlichen Vereins in Hamburg,* **21/22,** 73–85.

Enghoff, H. (1978*b*). Parthenogenesis and bisexuality in the millipede *Nemasoma varicorne* C. L. Koch 1847 (Diplopoda: Nemasomatidae). II. Distribution, substrate and abundance of the bisexual and thelytokus forms in some Danish forests. *Entomologiske Meddelelser,* **46,** 73–9.

Enghoff, H. (1979*a*). Taxonomic significance of the mandibles in the millipede Order Julida. In *Myriapod biology,* (ed. M. Camatini), pp. 27–38. Academic Press, London.

Enghoff, H. (1979*b*). The millipede genus *Okeanobates* (Diplopoda, Julida: Nemasomatidae). *Steenstrupia,* **5,** 161–78.

Enghoff, H. (1981). A cladistic analysis and classification of the millipede Order Julida. *Zeitschrift für Zoologische Systematik und Evolutionsforschung,* **19,** 285–319.

Enghoff, H. (1982*a*). The millipede genus *Cylindroiulus* on Madeira—an insular species swarm (Diplopoda, Julida: Julidae). *Entomologica Scandinavica,* Supplement 18, 1–142.

Enghoff, H. (1982*b*). An extraordinary new genus of the millipede family Nemasomatidae (Diplopoda: Julida). *Myriapodologica,* **1,** 69–80.

Enghoff, H. (1982*c*). The Zosteractinidae, a Nearctic family of millipedes (Diplopoda, Julida). *Entomologica Scandinavica,* **13,** 403–13.

Enghoff, H. (1983*a*). Adaptive radiation of the millipede genus *Cylindroiulus* on Madeira: habitat, body size and morphology (Diplopoda, Julida, Julidae). *Revue d'Ecologie et de Biologie du Sol,* **20,** 403–15.

Enghoff, H. (1983*b*). *Acipes*—a Macronesian genus of millipedes (Diplopoda, Julida, Blaniulidae). *Steenstrupia,* **9,** 137–79.

Enghoff, H. (1984*a*). Phylogeny of millipedes—a cladistic analysis. *Zeitschrift für Zoologische Systematik und Evolutionsforschung,* **22,** 8–26.

Enghoff, H. (1984*b*). Revision of the millipede genus *Choneiulus* (Diplopoda, Julida, Blaniulidae). *Steenstrupia,* **10,** 193–203.

Enghoff, H. (1985). Modified mouthparts in hydrophilous cave millipedes (Diplopoda). *Bijdragen tot de Dierkunde,* **55,** 67–77.

Enghoff, H. (1986). Leg polymorphism in a julid millipede, *Anaulaciulus inaequipes* n. sp. with a list of congeneric species (Diplopoda, Julida, Julidae). *Steenstrupia,* **12,** 117–25.

Enghoff, H. (1990). The ground-plan of chilognathan millipedes (external morphology). In *Proceedings of the 7th International Congress of Myriapodology,* (ed. A. Minelli), pp. 1–24. E. J. Brill, Leiden.

Evans, M. E. G. and Blower, J. G. (1973). A jumping millipede. *Nature,* **246,** 427–8.

Fain, A. (1987*a*). Notes on mites associated with Myriapoda. I. Three new astigmatid mites from afrotropical Myriapoda (Acari: Astigmata). *Bulletin de l'Institut Royal des Sciences Naturelles de Belgique (Entomologie)*, **57**, 161–72.

Fain, A. (1987*b*). Notes on mites associated with Myriapoda. II. Four new species of genus *Julolaelaps* Berlese 1916 (Acari: Laelapidae). *Bulletin de l'Institut Royal des Sciences Naturelles de Belgique (Entomologie)*, **57**, 203–8.

Fain, A. (1988). Notes on mites associated with Myriapoda. III. Two new species of the genus *Heterozercon* Berlese 1888 (Acari, Mesostigmata) from Afrotropical millipedes. *Bulletin et Annales de la Societé Royale Entomologique de Belgique*, **124**, 7–9.

Fairhurst, C. P. (1974). The adaptive significance of variations in the life cycle of Schizophylline millipedes. *Symposia of the Zoological Society of London*, **32**, 575–87.

Fairhurst, C. P. and Armitage, M. L. (1979). The British Myriapod Survey 1978. In *Myriapod biology*, (ed. M. Camatini), pp. 183–94. Academic Press, London.

Fairhurst, C. P., Barber, A. D., and Armitage, M. L. (1978). The British Myriapod Survey—April 1975. *Abhandlungen und Verhandlungen des Naturwissenschaftlichen Vereins in Hamburg*, **21/22**, 129–34.

Farquharson, P. A. (1974*a*). A study of the Malpighian tubules of the pill millipede *Glomeris marginata* (Villers). I. The isolation of the tubules in a Ringer solution. *Journal of Experimental Biology*, **60**, 13–28.

Farquharson, P. A. (1974*b*). A study of the Malpighian tubules of the pill millipede *Glomeris marginata* (Villers). II. The effect of variations in osmotic pressure and sodium and potassium concentrations on fluid production. *Journal of Experimental Biology*, **60**, 29–39.

Farquharson, P. A. (1974*c*). A study of the Malpighian tubules of the pill millipede *Glomeris marginata* (Villers). III. The permeability characteristics of the tubule. *Journal of Experimental Biology*, **60**, 41–51.

Fechter, H. (1961). Anatomie und Funktion der Kopfmuskulatur von *Cylindroiulus teutonicus* (Pocock). *Zoologische Jahrbücher (Anatomie)*, **79**, 479–528.

Federici, B. A. (1984.) Diseases of terrestrial isopods. *Symposia of the Zoological Society of London*, **53**, 233–45.

Fleming, W. E. and Hawley, I. M. (1950). A large scale test with DDT to control the Japanese beetle. *Journal of Economic Entomology*, **43**, 585–90.

Fontanetti, C. S. (1988). Histological studies in the testes of three Brazilian species of Diplopoda. *Journal of Advanced Zoology*, **9**, 87–91.

Fussey, G. D. and Varndell, I. M. (1980). The identification of the bisexual form of the bristly millipede *Polyxenus lagurus* (L. 1758) (Diplopoda: Polyxenida) at three coastal sites in England and Wales, using sex ratios. *Naturalist*, **105**, 151–4.

Gabe, M. (1954). Emplacement et connexions des cellules neurosécrétices chez quelques Diplopodes. *Compte Rendu Hebdomadaire des Séances de l'Académie des Sciences, Paris*, **239**, 828–30.

Gaffal, K. P., Tichy, H., Theiss, J., and Seelinger, G. (1975). Structural polarities in mechanosensitive sensilla and their influence on stimulus transmission (Arthropoda). *Zoomorphologie*, **82**, 79–103.

Geoffroy, J. J. (1981*a*). Étude d'un écosystème forestier mixte. V. Traits généraux du peuplement de Diplopodes édaphiques. *Revue d'Ecologie et de Biologie du Sol*, **18**, 357–72.

Geoffroy, J. J. (1981*b*). Modalités de la coexistence de deux Diplopodes, *Cylindroiulus punctatus* (Leach) et *Cylindroiulus nitidus* (Verhoeff) dans un écosystème forestier du Bassin Parisien. *Acta Oecologia Oecologia Generalis,* **2,** 357–72.

Geoffroy, J. J. (1984). *Opisthocheiron canayerensis* (Diplopoda: Craspedosomida): repartition de l'espèce et variation de la pigmentation. *Mémoires de Biospéology,* **11,** 295–302.

Geoffroy, J. J. (1990). La faune des diplopodes de France: un bilan des espèces. In *Proceedings of the 7th International Congress of Myriapodology,* (ed. A. Minelli), pp. 345–59. E. J. Brill, Leiden.

Geoffroy, J. J., Celerier, M. L., Garay, I., Rherissi, S., and Blandin, P. (1987). Approche quantitative des fonctions de transformation de la matère organique par les macroarthropodes saprophages (Isopoda et Diplopodes) dans un sol forestier à moder. Protocoles experimentaux et premiers résultats. *Revue d'Ecologie et de Biologie du Sol,* **24,** 573–90.

Gere, G. (1956). The examination of the feeding biology and the humificative function of Diplopoda and Isopoda. *Acta Biologica (Hungaricae),* **6,** 257–71.

Gibbs, A. J. (1952). A sporozoan blood parasite of *Archiulus moreleti* (Diplopoda). *Parasitology,* **42,** 74–6.

Gillon, Y. and Gillon, D. (1976). Comparaison par piégeage des populations de Diplopodes Iuliformes en zone de végétation naturelle et Champ d'Arachide. *Cahiers de l'Office de la Recherche Scientifique et Technique Outre-Mer, Série Biologie,* **11,** 121–7.

Gilyarov, M. S. (1970). Litter-destroying invertebrates and ways of increasing their useful activity. *Soviet Journal of Ecology,* **2,** 99–109.

Gilyarov, M. S. (1979). Soil fauna of brown soil in the Caucasus beech and fir mixed forests and some other communities. *Pedobiologia,* **19,** 408–24.

Gist, C. S. and Crossley, D. A. (1975). Feeding rates of some cryptozoa as determined by isotopic half-life studies. *Environmental Entomology,* **4,** 625–31.

Golovatch, S. I. (1990). On the distribution and faunogenesis of Crimean millipedes (Diplopoda). In *Proceedings of the 7th International Congress of Myriapodology,* (ed. A. Minelli), pp. 361–6. E. J. Brill, Leiden.

Gromysz-Kalkowska, K. (1970). The influence of body weight, external temperature, seasons of the year and fasting on respiratory metabolism in *Polydesmus complanatus* L. (Diplopoda). *Folia Biologica (Kraków),* **18,** 311–26.

Gromysz-Kalkowska, K. (1974). The effect of some exogenous factors and body weight on oxygen consumption in *Glomeris connexa* C. L. Koch (Diplopoda). *Folia Biologica (Kraków),* **22,** 37–49.

Gromysz-Kalkowska, K. (1976*a*). Influence of temperature and duration of acclimation on the metabolic rate in *Orthomorpha gracilis* (C. L. Koch) (Diplopoda). *Folia Biologica (Kraków),* **24,** 55–64.

Gromysz-Kalkowska, K. (1976*b*). The oxygen consumption of *Polyzonium germanicum* Brdt. (Diplopoda) in relation to some exogenous and endogenous factors. *Folia Biologica (Kraków),* **24,** 401–15.

Gromysz-Kalkowska, K. (1979). Some properties of the respiratory metabolism of the adult *Leptoiulus proximus proximus* (Nem.) (Diplopoda). *Folia Biologica (Kraków),* **27,** 129–45.

Gromysz-Kalkowska, K. (1980). Oxygen consumption of some species of millipedes (Diplopoda) under fasting conditions (in Polish). *Folia Societatis Scientiarum Lublinensis,* **22,** 41–8.

Gromysz-Kalkowska, K. and Stojalowska, W. (1983). Effect of temperature on respiration metabolism of two starved species of millipedes (Diplopoda). *Annales Universitatis Mariae Curie-Sklodowska,* **38C,** 183–91.

Gromysz-Kalkowska, K. and Tracz, H. (1983). Effect of temperature, food kind and body weight on the oxygen consumption by *Proteroiulus fuscus* (Am Stein) (Diplopoda, Blaniulidae). *Annals of Warsaw Agricultural University, Forestry and Wood Technology,* **30,** 35–42.

Gromysz-Kalkowska, K., Tracz, H., and Szubartowska, E. (1986). Some aspects of the respiration rate in *Ommatoiulus sabulosus* (Diplopoda). *Folia Societatis Scientiarum Lublinensis,* **22,** 65–76.

Gupta, A. P. (1979). Origin and affinities of Myriapoda. In *Myriapod biology,* (ed. M. Camatini), pp. 373–90. Academic Press, London.

Haacker, U. (1968). Deskriptive experimentelle und vergleichende untersuchungen zur Autokologie rhein-mainischer Diplopoden. *Oecologia,* **1,** 87–129.

Haacker, U. (1969). An attractive secretion in the mating behaviour of a millipede. *Zeitschrift für Tierpsychologie,* **26,** 988–90.

Haacker, U. (1974). Patterns of communication in courtship and mating behaviour of millipedes (Diplopoda). *Symposia of the Zoological Society of London,* **32,** 317–28.

Haacker, U. and Fuchs, S. (1970). Das paarungsverhalten von *Cylindroiulus punctatus* Leach. *Zeitschrift für Tierpsychologie,* **27,** 641–8.

Haacker, U. and Fuchs, S. (1972). Tree climbing in Pill-millipedes. *Oecologia,* **10,** 191–2.

Halkka, R. (1958). Life history of *Schizophyllum sabulosum* (L.) (Diplopoda, Iulidae). *Annales Zoologici Societatis Zoologicae-Botanicae Fennicae (Vanamo),* **19,** 1–72.

Hall, F. R., Hollingworth, R. M., and Shankland, D. L. (1969). Cyanide tolerance in millipedes: comparison of respiration in millipedes and insects. *Entomological News,* **80,** 277–82.

Hall, F. R., Hollingworth, R. M., and Shankland, D. L. (1971). Cyanide tolerance in millipedes: the biochemical basis. *Comparative Biochemistry and Physiology,* **38B**, 723–37.

Halstead, B. W. and Ryckman, R. (1949). Injurious effects from contacts with millipedes. *Medical Arts and Sciences,* **3,** 16–18.

Haneveld, G. T. (1958). Eye lesions caused by the exudate of tropical millipedes. I. Report on a case. *Tropical and Geographical Medicine,* **10,** 165–7.

Hanlon, R. D. G. (1981*a*). Some factors influencing microbial growth on soil animal faeces. 1. Bacterial and fungal growth on particulate oak leaf litter. *Pedobiologia,* **21,** 257–63.

Hanlon, R. D. G. (1981*b*). Some factors influencing microbial growth on soil animal faeces. 2. Bacterial and fungal growth on soil animal faeces. *Pedobiologia,* **21,** 264–70.

Hanlon, R. D. G. and Anderson, J. M. (1980). The influence of macroarthropod feeding activities on microflora in decomposing oak leaves. *Soil Biology and Biochemistry,* **12,** 255–61.

Hannibal, J. (1981). The spinose archipolypods: giant millipedes of the Coal Age. *The Explorer,* **23,** 15–17.

Hannibal, J. (1984). Pill millipedes from the Coal Age. *Field Museum of Natural History Bulletin,* **55**(8), 12–16.

Hannibal, J. T. (1986). The enduring myriapods. *Earth Science,* **39,** 21–3.

Hannibal, J. T. and Feldmann, R. M. (1981). Systematics and functional morphology of Oniscomorph millipedes (Arthropoda: Diplopoda) from the Carboniferous of North America. *Journal of Paleontology,* **55,** 730–46.

Hannibal, J. and Talerico, C. (1986). Millipede hording. A curious phenomenon of nature. *Field Museum of Natural History Bulletin,* **57,** 24–5.

Harborne, J. B. (1988). *Introduction to ecological biochemistry,* (3rd edn). Academic Press, London.

Hartenstein, R. (1982). Soil macroinvertebrates, aldehyde, oxidase, catalase, cellulase and peroxidase. *Soil Biology and Biochemistry,* **14,** 387–91.

Hassall, M. and Rushton, S. P. (1982). The role of coprophagy in the feeding strategies of terrestrial isopods. *Oecologia,* **53,** 374–81.

Hassall, M. and Rushton, S. P. (1985). The adaptive significance of coprophagous behaviour in the terrestrial isopod *Porcellio scaber. Pedobiologia,* **28,** 169–75.

Hastings, J. W. and Davenport, D. (1957). The luminescence of the millipede *Luminodesmus sequoiae. Biological Bulletin,* **113,** 120–8.

Haupt, J. (1979). Phylogenetic aspects of recent studies on myriapod sense organs. In *Myriapod biology,* (ed. M. Camatini), pp. 391–406. Academic Press, London.

Heath, G. W., Arnold, M. K., and Edwards, C. A. (1966). Studies in leaf litter breakdown. I. Breakdown rates of leaves of different species. *Pedobiologia,* **6,** 1–12.

Heath, J., Bocock, K. L., and Mountford, M. D. (1974). The life history of the millipede *Glomeris marginata* (Villers) in North-West England. *Symposia of the Zoological Society of London,* **32,** 433–62.

Helb, H. W. (1975). Zum Massenauftrefen des Schurfüssers *Schizophyllum sabulosum* (Myriapoda: Diplopoda). *Entomologica Germanica,* **1,** 376–81.

Hicking, W. (1979). Kristalline Einlagerungen in der Kutikula von *Tachypodoiulus niger* (Leach) (Diplopoda, Myriapoda). *Biomineralisation,* **10,** 66–9.

Hoffman, R. L. (1965). A new species in the diplopod genus *Archispirostreptus* from southwestern Arabia. *Entomologische Mitteilungen aus dem Zoologischen Staatsintitut und Zoologischen Museum, Hamburg,* **3** (51), 17–23.

Hoffman, R. L. (1969). Myriapoda, exclusive of Insecta. In *Treatise on Invertebrate Paleontology,* (ed. R. Moore), pp. R572–606. Geological Society of America, Lawrence, Kansas.

Hoffman, R. L. (1978). On the classification and phylogeny of Chelodesmoid Diplopoda. *Abhandlungen und Verhandlungen des Naturwissenschaftlichen Vereins in Hamburg,* **21/22,** 129–34.

Hoffman, R. L. (1979). *Classification of the Diplopoda.* Museum d'Histoire Naturelle, Genève.

Hoffman, R. L. (1982). Diplopoda. In *Synopsis and classification of living organisms,* (ed. S. P. Parker), pp. 689–724. McGraw-Hill, New York.

Hoffman, R. L. and Payne, J. A. (1969). Diplopods as carnivores. *Ecology,* **50,** 1096–8.

Hölldobler, B. and Wilson, E. O. (1990). *The ants.* Springer-Verlag, Berlin.

Hopkin, S. P. (1989). *Ecophysiology of metals in terrestrial invertebrates.* Elsevier Applied Science, Barking.

Hopkin, S. P., Watson, K., Martin, M. H., and Mould, M. L. (1985). The assimilation of heavy metals by *Lithobius variegatus* and *Glomeris marginata* (Chilopoda; Diplopoda). *Bijdragen tot de Dierkunde,* **55,** 88–94.

Horst, D. J. van der and Oudejans, R. C. H. M. (1973). Cyclopropane fatty acids in the desert millipede *Orthoporus ornatus* (Girard) (Myriapoda: Diplopoda: Spirostreptida). *Comparative Biochemistry and Physiology,* **46B,** 277–81.

Horst, D. J. van der and Oudejans, R. C. H. M. (1978). Cyclopropane fatty acids in millipedes: their puzzling function. *Abhandlungen und Verhandlungen des Naturwissenschaftlichen Vereins in Hamburg,* **21/22,** 349–52.

Horst, D. J. van der, Oudejans, R. C. H. M., and Zandee, D. I. (1972). Occurrence of cyclopropane fatty acids in females and eggs of the millipede *Graphidostreptus tumuliporus* (Karsch) (Myriapoda, Diplopoda), as contrasted with their absence in the males. *Comparative Biochemistry and Physiology,* **41B,** 417–23.

Horst, D. J. van der, Oudejans, R. C. H. M., Plug, A. G., and Harmelen, H. J. M. van (1973). Biosynthesis of cyclopropane fatty acids in the millipede *Graphidostreptus tumuliporus* (Karsch), (Diplopoda: Spirostreptida). *Comparative Biochemistry and Physiology,* **46B,** 395–404.

Hubert, M. (1972). Les tubes de Malpighi de *Cylindroiulus teutonicus* Pocock (*londinensis* C. L. K.) (Dipopode, Iuloidea). Données histologiques et infrastructurales. *Compte Rendu Hebdomadaire des Séances de l'Académie des Sciences, Paris,* **275D,** 1043–6.

Hubert, M. (1973). Les reins labiaux de *Cylindroiulus teutonicus* Pocock (*londinensis* C. L. K.) (Diplopode, Iuloidea). Étude histologique et ultrastructurale. *Compte Rendu Hebdomadaire des Séances de l'Académie des Sciences, Paris,* **277D,** 1507–9.

Hubert, M. (1974). Le tissu adipeux de *Cylindroiulus teutonicus* Pocock (*londinensis* C. L. K.) Diplopode, Iuloidea; étude histologique et ultrastructurale. *Compte Rendu Hebdomadaire des Séances de l'Académie des Sciences, Paris,* **278D,** 3343–6.

Hubert, M. (1975). Sur la nature des accumulations minerales et puriques chez *Cylindroiulus teutonicus* Pocock (*londinensis* C. L. K., Diplopode, Iuloidea). *Compte Rendu Hebdomadaire des Séances de l'Académie des Sciences, Paris,* **281D,** 151–4.

Hubert, M. (1977). Contribution à l'étude des organs excreteurs et de l'excrétion chez les Diplopodes. Thèse de Doctorat d'État, Université de Rennes, France.

Hubert, M. (1978*a*). Données histophysiologiques complémentaires sur les bioaccumulations minérales et puriques chez *Cylindroiulus londinensis* (Leach, 1814) (Diplopode, Iuloidea). *Archives de Zoologie Experimentale et Générale,* **119,** 669–83.

Hubert, M. (1978*b*). Les cellules hépatiques de *Cylindroiulus londinensis* (Leach, 1814) (Diplopode, Iuloidea). *Compte Rendu Hebdomadaire des Séances de l'Académie des Sciences, Paris,* **286D,** 627–30.

Hubert, M. (1979*a*). Localization and identification of mineral elements and nitrogenous waste in Diplopoda. In *Myriapod biology,* (ed. M. Camatini), pp. 127–34. Academic Press, London.

Hubert, M. (1979*b*). L'intestin moyen de *Cylindroiulus londinensis* Leach (*psylopygus* Latzel) (Diplopode, Iuloidea): observations ultrastructurales en relation avec la fonction d'accumulation. *Compte Rendu Hebdomadaire des Séances de l'Académie des Sciences, Paris,* **289D,** 749–52.

Hubert, M. (1981). La cuticle proctodéale de *Cylindroiulus londinensis* Leach *(psylopygus* Latzel) (Diplopode, Iuloidea): étude ultrastructurale des dépressions épicuticulaires. *Bulletin du Muséum National d'Histoire Naturelle, Série IV,* **3A,** 815–23.

Hubert, M. (1988). Le complexe anatomique et fonctionnel cellules hepatiques—mesenteron de *Cylindroiulus londinensis* Leach (*psylopygus* Latzel): étude ultrastructurale et spectrographique. *Bulletin de la Société Zoologique de France,* **113,** 191–8.

Hubert, M. and Razet, P. (1965). Sur les principaux éléments du catabolisme azoté chez les myriapodes. *Compte Rendu Hebdomadaire des Séances de l'Académie des Sciences, Paris,* **261,** 797–800.

Iatrou, G. D. and Stamou, G. P. (1988). Post embronic growth of *Glomeris balcanica* (Diplopoda, Glomeridae). *Pedobiologia,* **32,** 343–53.

Iatrou, G. D. and Stamou, G. P. (1989). Seasonal activity patterns of *Glomeris balcanica* (Diplopoda, Glomeridae) in an evergreen-sclerophyllous formation in northern Greece. *Revue d'Ecologie et de Biologie du Sol,* **26,** 491–503.

Iatrou, G. D. and Stamou, G. P. (1990*a*). Studies on the life cycle of *Glomeris balcanica* (Diplopoda, Glomeridae) under laboratory conditions. *Pedobiologia,* **34,** 173–81.

Iatrou, G. D. and Stamou, G. P. (1990*b*). Growth patterns of *Glomeris balcanica* (Diplopoda, Glomeridae). In *Proceedings of the 7th International Congress of Myriapodology* (ed. A. Minelli), pp. 337–44. E. J. Brill, Leiden.

Iatrou, G. D. and Stamou, G. P. (1991). The life cycle and spatial distribution of *Glomeris balcanica* (Diplopoda, Glomeridae) in an evergreen-sclerophyllous formation in northern Greece. *Pedobiologia,* **35,** 1–10.

Ineson, P. and Anderson, J. M. (1985). Aerobically isolated bacteria associated with the gut and faeces of the litter feeding macroarthropods *Oniscus asellus* and *Glomeris marginata. Soil Biology and Biochemistry,* **17,** 843–9.

Ishii, K. (1988). On the significance of the mandible as a diagnostic character in the taxonomy of Penicillate Diplopods (Diplopoda, Polyxenidae). *Canadian Entomologist,* **120,** 955–63.

Ishii, K. and Yamaoka, H. (1982). The species and number of symbiotic Penicillate millipeds in arboreal ant nests. *Canadian Entomologist,* **114,** 767–8.

IUCN (1983). *Invertebrate red data book.* International Union for Conservation of Nature and Natural Resources, Gland, Switzerland.

Jackson, R. M. and Raw, F. (1966). *Life in the Soil.* Edward Arnold, London.

Jamieson, B. G. M. (1987). *The Ultrastructure and phylogeny of insect spermatozoa.* Cambridge University Press.

Jeekel, C. A. W. (1970). Nomenclator generum et familiarum Diplopodorum: a list of the genus and family-group names in the Class Diplopoda from the 10th edition of Linnaeus 1758 to the end of 1957. *Monografieen van de Nederlandsche Entomologische Vereeniging,* Number 5.

Jeekel, C. A. W. (1974). The group taxonomy and geography of the Sphaerotherida (Diplopoda). *Symposia of the Zoological Society of London,* **32,** 41–52.

Jeekel, C. A. W. (1985*a*) (ed.). Proceedings of the 6th International Congress of Myriapodology. *Bijdragen tot de Dierkunde,* **55,** 1–218.

Jeekel, C. A. W. (1985*b*). The distribution of the diplocheta and the 'lost' continent Pacifica (Diplopoda). *Bijdragen tot de Dierkunde,* **55,** 100–12.

Jeram, A. J., Selden, P. A., and Edwards, D. (1990). Land animals in the Silurian: arachnids and myriapods from Shropshire, England. *Science,* **250,** 658–61.

Jocteur Monrozier, L. and Robin, A. M. (1988). Action de la fauna du sol sur une litière de feuillu: application de techniques pyrolytiques à l'étude des modifications subies par les feuilles de charme (*Carpinus betulus*) ingérées par *Glomeris marginata. Revue d'Ecologie et de Biologie du Sol,* **25,** 203–14.

Johns, P. M. (1962). Introduction to the endemic and introduced millipedes of New Zealand. *New Zealand Entomologist,* **3,** 38–46.

Johns, P. M. (1979). Speciation in New Zealand Diplopoda. In *Myriapod biology,* (ed. M. Camatini), pp. 49–57. Academic Press, London.

Johnson, I. T. and Riegel, J. A. (1977*a*). Ultrastructural tracer studies on the permeability of the Malpighian tubule of the pill millipede *Glomeris marginata* (Villers). *Cell and Tissue Research,* **182,** 549–56.

Johnson, I. T. and Riegel, J. A. (1977*b*). Ultrastructural studies on the Malpighian tubule of the pill millipede *Glomeris marginata* (Villers) (Myriapoda, Diplopoda). General morphology and localisation of phosphatase enzymes. *Cell and Tissue Research,* **180,** 357–66.

Jolivet, P. (1986). Le millepatte de Santo Antao (Iles du Cap Vert) ou comment une espèce inoffensive peut devenir un ravageur! *Entomologiste,* **42,** 45–56.

Joly, R. and Descamps, M. (1987). Histology and ultrastructure of the Myriapod brain. In *Arthropod brain,* (ed. A. P. Gupta), pp. 135–57. Wiley, New York.

Joly, R. and Descamps, M. (1988). Endocrinology of myriapods. In *Endocrinology of selected invertebrate types: invertebrate endocrinology Volume 2,* (ed. H. Laufer and R. G. H. Downer), pp. 429–49. Liss, New York.

Juberthie, C. and Juberthie-Jupeau, L. (1974). Étude ultrastructurale de l'organe neurohémal cérébral de *Speleoglomeris doderoi* Silvestri, Myriapode Diplopode cavernicole. *Symposia of the Zoological Society of London,* **32,** 199–210.

Juberthie-Jupeau, L. (1976*a*). Existence d'organes neuraux intracérébraux chez les Glomeridia (Diplopodes) épigés et cavernicoles. *Compte Rendu Hebdomadaire des Séances de l'Académie des Sciences, Paris,* **264D,** 89–92.

Juberthie-Jupeau, L. (1967*b*). Données sur le système endocrinien de quelques Diplopodes Oniscomorphes (Myriapodes). *Compte Rendu Hebdomadaire des Séances de l'Académie des Sciences, Paris,* **265D,** 1527–9.

Juberthie-Jupeau, L. (1973). Étude ultrastructurale des corps paraoesophagiens chez un Diplopode Oniscomorphe *Loboglomeris pyrenaica* Latzel. *Compte Rendu Hebdomadaire des Séances de l'Académie des Sciences, Paris,* **276D,** 169–72.

Juberthie-Jupeau, L. (1974). Action de la température sur le développement embryonnaire de *Glomeris marginata* (Villers). *Symposia of the Zoological Society of London,* **32,** 289–300.

Juberthie-Jupeau, L. (1976). Fine structure of postgonopodial glands of a myriapod *Glomeris marginata* (Villers). *Tissue and Cell,* **8,** 293–304.

Juberthie-Jupeau, L. (1983). Neurosecretory systems and neurohaemal organs of Myriapoda. In *Neurohaemal organs of arthropods,* (ed. A. P. Gupta), pp. 204–78. C. C. Thomas, Springfield, Illinois.

Juberthie-Jupeau, L. and Tabacaru, I. (1968). Glandes postgonopodiales des Oniscomorphes (Diplopodes, Myriapodes). *Revue d'Ecologie et de Biologie du Sol,* **4,** 605–18.

Kanaka, R. and Chowdaiah, B. N. (1974). Studies on male reproductive pattern in some Indian Diplopoda. *Symposia of the Zoological Society of London,* **32,** 261–272.

Kaplan, D. L. and Hartenstein, R. (1978). Studies on monoxygenases and dioxygenases in soil macroinvertebrates and bacterial isolates from the gut of the terrestrial isopod *Oniscus asellus* L. *Comparative Biochemistry and Physiology,* **60B,** 47–50.

Karamaouna, M. (1987). Ecology of millipedes in mediterranean coniferous ecosystems of Southern Greece. Unpublished Ph.D. thesis, University of Athens.

Karamaouna, M. (1990). Aspects of ecology of *Polyxenus lagurus* in Mediterranean conifer formations of Greece (Diplopoda: Penicillata). In *Proceedings of the 7th International Congress of Myriapodology,* (ed. A. Minelli), pp. 255–64. E. J. Brill, Leiden.

Karamaouna, M. and Geoffroy, J. J. (1985). Millipedes of a maquis ecosystem (Naxos Island, Greece): preliminary description of the population (Diplopoda). *Bijdragen tot de Dierkunde,* **55,** 113–15.

Kayed, A. N. (1978). Consumption and assimilation of food by *Ophyiulus pilosus* (Newport). *Abhandlungen und Verhandlungen des Naturwissenschaftlichen Vereins in Hamburg,* **21/22,** 115–20.

Kayed, A. N. (1986*a*). Comparative studies on the methods used in estimating food consumption in diplopods. *Proceedings of the Zoological Society of the United Arab Republic,* **12,** 169–87.

Kayed, A. N. (1986*b*). Comparison between the assimilation quotient and the consumption quotient in the millipede *Orthomorpha gracilis. Proceedings of the Zoological Society of the United Arab Republic,* **12,** 189–97.

Kennedy, G. Y. (1978). Pigments of the Myriapoda. *Abhandlungen und Verhandlungen des Naturwissenschaftlichen Vereins in Hamburg,* **21/22,** 365–71.

Kethley, J. B. (1974). Developmental chaetotaxy of a paedomorphic celaenopsoid *Neotenogynium malkini* n.g., n.sp. (Acari: Parasitiformes: Neotenogyniidae, n.fam.) associated with millipedes. *Annals of the Entomological Society of America,* **67,** 571–9.

Kheirallah, A. M. (1978). The consumption and utilization of two different species of leaf litter by a laboratory population of *Orthomorpha gracilis* (Diplopoda, Polydesmoidea). *Entomologia Experimentalis et Applicata,* **23,** 14–19.

Kheirallah, A. M. (1979). Behavioural preference of *Julus scandinavius* (Myriapoda) to different species of leaf litter. *Oikos,* **33,** 466–71.

Kheirallah, A. M. and Shabana, M. B. (1975). Effect of food qualities on protein pattern of the haemolymph of a millipede *Orthomorpha gracilis. Entomologia Experimentalis et Applicata,* **18,** 423–8.

Kime, R. D. (1990*a*). *A provisional atlas of european myriapods Part 1.* European Invertebrate Survey, Luxembourg.

Kime, R. D. (1990*b*). Spatio-temporal distribution of European millipedes. In *Proceedings of the 7th International Congress of Myriapodology,* (ed. A. Minelli), pp. 367–80. E. J. Brill, Leiden.

Kime, R. D. (1991). On the abundance of West-European millipedes (Diplopoda). *Proceedings of the 8th International Congress of Myriapodology. Veröffentlichungen der Universität Innsbruck.* (In press.)

Kime, R. D. and Wauthy, G. (1984). Aspects of relationships between millipedes, soil texture and temperature in deciduous forests. *Pedobiologia,* **26,** 387–402.

Kime, R. D., Wauthy, G., Delecour, F., and Dufrene, M. (1991). Distribution spatiale et préférences écologique chez les Diplopodes du sol. *Mémoires de la Société Royale Entomologie de Belgique,* **35.** (In press.)

Koch, C. L. (1863). *Die Myriapoden: Getrau Nach der Natur Abgebildet und Beschrieben,* 2 Vols. Halle.

Koch, L. E. (1985). Pin cushion millipedes (Diplopoda: Polyxenida): their aggregations and identity in Western Australia. *Western Australian Naturalist,* **16,** 30–32.

Köhler, H. R. and Alberti, G. (1990). Morphology of the mandibles in the millipedes (Diplopoda, Arthropoda). *Zoologica Scripta,* **19,** 195–202.

Köhler, H. R. and Alberti, G. (1991). The effect of heavy metal stress of the intestine of diplopods. *Proceedings of the 8th International Congress of Myriapodology. Veröffentlichungen der Universität Innsbruck.* (In press.)

Köhler, H. R., Ullrich, B., Storch, V., Schairer, H. V., and Alberti, G. (1989). Massen- und Energiefluss bei Diplopoden und Isopoden. *Mitteilungen der Deutschen Gesellschaft für Allgemeine und Angewandte Entomologie,* **7,** 263–8.

Köhler, H. R., Alberti, G., and Storch, V. (1991). The influence of the mandibles of Diplopoda on the food—a dependence of fine structure and assimilation efficiency. *Pedobiologia,* **35,** 108–16.

Korsós, Z. (1990). Computerized database and mapping of Myriapods in Hungary. In *Proceedings of the 7th International Congress of Myriapodology,* (ed. A. Minelli), pp. 381–3. E. J. Brill, Leiden.

Krabbe, E. (1979). The first pair of legs in male Spirostreptidae: their function and taxonomic importance. In *Myriapod biology,* (ed. M. Camatini), pp. 59–72. Academic Press, London.

Kraus, O. (1974). On the morphology of Palaezoic Diplopods. *Symposia of the Zoological Society of London,* **32,** 13–22.

Kraus, O. (1978*a*) (ed.). Proceedings of the Third International Congress of Myriapodology. *Abhandlungen und Verhandlungen des Naturwissenschaftlichen Vereins in Hamburg,* **21/22.**

Kraus, O. (1978*b*). Zoogeography and plate tectonics. Introduction to a general discussion. *Abhandlungen und Verhandlungen des Naturwissenschaftlichen Vereins in Hamburg,* **21/22,** 33–41.

Kraus, O. (1990). On the so-called thoracic segments in Diplopoda. In *Proceedings of the 7th International Congress of Myriapodology,* (ed. A. Minelli), pp. 63–8. E. J. Brill, Leiden.

Krishnan, G. and Ravindranath, M. H. (1973). Phenol oxidase in the blood cells of millipedes. *Journal of Insect Physiology,* **19,** 647–53.

Krishnan Nair, V. S. and Prabhu, V. K. K. (1971). On the free amino acids in the haemolymph of a millipede. *Comparative Biochemistry and Physiology,* **38B,** 1–4.

Krivoluckij, D. A., Tichomirova, A. L., and Turcaninova, V. A. (1972). Strukturaenderungen des Tierbesatzes (Land—und Bodenwirbellose) unter dem Einfluss der Kontamination des Bodens mit Sr^{90}. *Pedobiologia,* **12,** 374–80.

Krivolutskii, D. A. and Filippova, N. F. (1979). Radiosensitivity of millipedes. *Radiobiologya,* **19,** 582–5.

Kubrakiewicz, J. (1989). Deposition of calcium salts in oocytes and ovarian somatic tissue of millipedes. *Tissue and Cell,* **21,** 443–6.

Kubrakiewicz, J. (1991*a*). Ultrastructural investigation of the ovary structure of *Ophyiulus pilosus* (Myriapoda, Diploda). *Zoomorphology,* **110,** 133–8.

Kubrakiewicz, J. (1991*b*). Ovary structure and oogenesis of *Polyxenus lagurus* (L.) (Diplopoda, Pselaphognatha). An ultrastructural study. *Zoologische Jahrbücher (Anatomie),* **121,** 81–93.

Kuhnelt, W. (1976). *Soil biology with special reference to the Animal Kingdom,* 2nd ed. (ed. N. Walker). Faber and Faber, London.

Kurnik, I. (1988). Zur Taxonomie ostalpiner Chordeumatida: Vulvanmorphologie und Identifikation der Weibchen. *Zoologische Jahrbücher (Systematik),* **115,** 229–68.

Kurnik, I. and Thaler, K. (1985). Die Vulven der Chordeumatida: Merkmale von Taxonomischer Bedeutung (Diplopoda; Helminthomorpha). *Bijdragen tot de Dierkunde,* **55,** 116–24.

Lanzavecchia, G. and Camatini, M. (1979). Phylogenetic problems and muscle cell ultrastructure in Onychophora. In *Myriapod biology,* (ed. M. Camatini), pp. 407–17. Academic Press, London.

Lauterbach, K. E. (1972). Über die sogenannte Ganzbein-Mandibel der Tracheaten, insbesondere der Myriaopoda. *Zoologischer Anzeiger,* **188,** 145–54.

Lawrence, R. F. (1952). Variation in the leg numbers of a South African millipede *Gymnostreptus pyrrocephalus* C. Koch. *Annals and Magazine of Natural History, Series 12,* **5,** 1044–51.

Levine, N. D. (1986). *Gibbsia archiuli* (Apicomplexa, Eucoccidiorida) n.g., n.sp., from the millipede *Archiulus moreleti. Journal of Protozoology,* **33,** 300–1.

Lewis, J. G. E. (1971*a*). The life history and ecology of the millipede *Tymbodesmus falcatus* (Polydesmida: Gomphodesmidae) in northern Nigeria with notes on *Sphenodesmus sheribongensis. Journal of Zoology,* **164,** 551–63.

Lewis, J. G. E. (1971*b*). The life history and ecology of three paradoxosomatid millipedes (Diplopoda: Polydesmida) in northern Nigeria. *Journal of Zoology,* **165,** 431–52.

Lewis, J. G. E. (1974). The ecology of centipedes and millipedes in Northern Nigeria. *Symposia of the Zoological Society of London,* **32,** 423–31.

Lewis, J. G. E. (1984). Notes on the biology of some common millipedes of the Gunung Mulu National Park, Sarawak, Borneo. *Sarawak Museum Journal,* **33,** 179–85.

Loomis, H. F. (1972). Millipedes associated with ants in Washington state. *Florida Entomologist,* **55,** 145–51.

Loomis, H. F. and Schmitt, R. (1971). The ecology, distribution and taxonomy of the millipedes of Montana west of the continental divide. *Northwest Science,* **45,** 107–31.

Lyford, W. H. (1943). Palatability of freshly fallen leaves of forest trees to millipedes. *Ecology,* **24,** 252–61.

McBrayer, J. F. (1973). Exploitation of deciduous leaf litter by *Apheloria montana* (Diplopoda: Eurydesmidae). *Pedobiologia,* **13,** 90–8.

McIver, S. B. (1975). Structure of cuticular mechanoreceptors of arthropods. *Annual Reviews of Entomology,* **20,** 381–97.

McKillup, S. C. (1988). Behaviour of the millipedes *Ommatoiulus moreleti, Ophyiulus verruculiger* and *Oncocladosoma castaneum* in response to visible light: an explanation for the invasion of houses by *Ommatoiulus moreleti. Journal of Zoology,* **215,** 35–46.

McKillup, S. C. and Bailey, P. T. (1990). Prospects for the biological control of the introduced millipede, *Ommatoiulus moreleti* (Lucas) (Julidae), in South Australia. In *Proceedings of the 7th International Congress of Myriapodology,* (ed. A. Minelli), pp. 265–70. E. J. Brill, Leiden.

McKillup, S. C., Allen, P. G., and Skewes, M. A. (1988). The natural decline of an introduced species following its initial increase in abundance; an explanation for *Ommatoiulus moreleti* in Australia. *Oecologia,* **77,** 339–42.

MAFF (1984). *Millipedes and Centipedes.* Ministry of Agriculture Fisheries and Food, Advisory Leaflet. HMSO, London.

Mangum, C. P., Scott, J. L., Black, R. E. L., Miller, K. I., and Van Holde, K. E. (1985). Centipedal hemocyanin: Its structure and its implications for arthropod phylogeny. *Proceedings of the National Academy of Sciences of the United States of America,* **82,** 3721–5.

Manier, J. M. and Boissin, L. (1978). Étude ultrastructurale comparative de la spermiogenèse des Diplopodes. *Abhandlungen und Verhandlungen des Naturwissenschaftlichen Vereins in Hamburg,* **21/22,** 197–202.

Mantel, L. H. (1979). Terrestrial invertebrates other than insects. In *Comparative physiology of osmoregulation in animals,* Volume 1, (ed. G. M. O. Maloiy), pp. 175–218. Academic Press, New York.

Manton, S. M. (1954). The evolution of arthropodan locomotory mechanisms. Part 4. The structure, habits and evolution of the Diplopoda. *Journal of the Linnean Society of London (Zoology),* **42,** 299–368.

Manton, S. M. (1956). The evolution of arthropod locomotory mechanisms. Part 5. The structure, habits and evolution of the Pselaphognatha (Diplopoda). *Journal of the Linnean Society of London (Zoology),* **43,** 153–87.

Manton, S. M. (1958). The evolution of arthropod locomotory mechanisms. Part 6. Habits and evolution of the Lysiopetaloidea (Diplopoda), some principles of leg design in Diplopoda and Chilopoda, and limb structure in Diplopoda. *Journal of the Linnean Society of London (Zoology),* **43,** 489–556.

Manton, S. M. (1961). The evolution of arthropodan locomotory mechanisms. Part 7. Functional requirements and body design in Colobognatha (Diplopoda), together with a comparative account of diplopod burrowing techniques, trunk musculature and segmentation. *Journal of the Linnean Society of London (Zoology),* **44,** 383–462.

Manton, S. M. (1964). Mandibular mechanisms and the evolution of arthropods. *Philosophical Transactions of the Royal Society of London, Series B,* **247,** 1–183.

Manton, S. M. (1966). The evolution of arthropod locomotory mechanisms. Part 9. Functional requirements and body design in Symphyla and Pauropoda and the relationships between Myriapoda and pterygote insects. *Journal of the Linnean Society of London (Zoology),* **46,** 103–41.

Manton, S. M. (1972). The evolution of arthropodan locomotory mechanisms. Part 10. Locomotory habits, morphology and evolution of the hexapod classes. *Journal of the Linnean Society of London (Zoology),* **51,** 203–400.

Manton, S. M. (1973). The evolution of arthropodan locomotory mechanisms. Part 11. Habits, morphology and evolution of the Uniramia (Onychophora, Myriapoda, Hexapoda) and comparisons with the Arachnida, together with a functional review of uniramian musculature. *Zoological Journal of the Linnean Society of London,* **53,** 257–375.

Manton, S. M. (1974). Segmentation in Symphyla, Chilopoda and Pauropoda in relation to phylogeny. *Symposia of the Zoological Society of London,* **32,** 163–90.

Manton, S. M. (1977). *The Arthropoda. Habits, functional morphology and evolution.* Clarendon Press, Oxford.

Marcus, E. R., Baransky, M., Kooker, de G. M., and Knox, C. M. (1987). Haemolymph composition of the Kalahari millipede *Triaenostreptus triodus* (Attems). *Comparative Biochemistry and Physiology,* **87A,** 603–6.

Marcuzzi, G. and Turchetto-Lafisca, M. L. (1977). On lipases in litter feeding invertebrates. *Pedobiologia,* **17,** 135–44.

Márialigeti, K., Jáger, K., Szabó, I. M., Pobozsny, M., and Dzingov, A. (1984). The faecal actinomycete flora of *Protracheoniscus amoenus* (Woodlice: Isopoda). *Acta Microbiologica Hungarica,* **31,** 339–44.

Márialigeti, K., Contreras, E., Barabas, G., Heydrich, M., and Szabó, I. M. (1985). True Actinomycetes of millipedes (Diplopoda). *Journal of Invertebrate Pathology,* **45,** 120–1.

Martin, J. S. and Kirkham, J. B. (1989). Dynamic role of microvilli in peritrophic membrane formation. *Tissue and Cell,* **21,** 627–38.

Mauriès, J. P. (1960). Note sur la répartition des Diplopodes dans le Massif du

Néouvieille (Hautes-Pyrénées). *Bulletin de la Société Zoologique de France,* **85,** 409–11.

Mauriès, J. P. (1969). Observations sur la biologie (sexualité, periodomorphose) de *Typholoblaniulus lorifer consoranensis* Brölemann (Diplopoda, Blaniulidae). *Annales de Speleologie,* **24,** 495–504.

Mauriès, J. P. (1974). Intérêt phylogénique et biogéographique de quelques diplopodes récemment décrits du Nord de l'Espagne. *Symposia of the Zoological Society of London,* **32,** 53–63.

Mauriès, J. P. (1987). Les Modes de peuplement des îles oceaniques par les Diplopodes: le cas des Antilles français. *Bulletin de la Société Zoologique de France,* **112,** 343–53.

Mead, M. and Gilhodes, J. C. (1974). Organisation temporelle de l'activité locomotrice chez un animal cavernicole *Blaniulus lichtensteini* Bröl. (Diplopoda). *Journal of Comparative Physiology,* **90,** 47–52.

Meidell, B. A. (1970). On the distribution, sex ratio, and development of *Polyxenus lagurus* (L.) in Norway. *Norsk Entomologisk Tidsskrift,* **17,** 147–52.

Meidell, B. (1979). Norwegian myriapods; some zoogeographical remarks. In *Myriapod biology,* (ed. M. Camatini), pp. 195–201. Academic Press, London.

Meinwald, Y. C., Meinwald, J., and Eisner, T. (1966). 1,2-dialkyl-4(3H)-quinazolinones in the defensive secretion of a millipede (*Glomeris marginata*). *Science,* **154,** 390–1.

Meinwald, J., Smolanoff, J., McPhail, A. T., Miller, R. W., Eisner, T., and Hicks, K. (1975). Nitropolyzonamine: a spirocyclic nitro compound from the defensive glands of a millipede (*Polyzonium rosalbum*). *Tetrahedron Letters,* **28,** 2367–70.

Meske, C. (1962). Untersuchungen zur Sinnesphysiologie von Diplopoden und Chilopoden. *Zeitschrift für Vergleichende Physiologie,* **45,** 61–77.

Messner, B. and Adis, J. (1988). Die Plastronstrukturen der bishereinzigen submers lebenden Diplodenart *Gonographis adisi* Hoffman 1985 (Pyrgodesmidae, Diplopoda). *Zoologische Jahrbücher (Anatomie),* **117,** 277–90.

Meyer, E. (1979). Life cycles and ecology of high alpine Nematophora. In *Myriapod biology,* (ed. M. Camatini), pp. 295–306. Academic Press, London.

Meyer, E. (1985). Distribution, activity, life-history and standing crop of Iulidae (Diplopoda, Myriapoda) in the Central High Alps (Tyrol, Austria). *Holarctic Ecology,* **8,** 141–50.

Meyer, E. (1990). Altitude-related changes of life histories of Chordeumatida in the Central Alps (Tyrol, Austria). In *Proceedings of the 7th International Congress of Myriapodology,* (ed. A. Minelli), pp. 311–22. E. J. Brill, Leiden.

Meyer, E. and Eisenbeis, G. (1985). Water relations in millipedes from some alpine habitat types (Central Alps, Tyrol) (Diplopoda). *Bijdragen tot de Dierkunde,* **55,** 131–42.

Miley, H. H. (1927). Development of the male gonopods and life history studies of a polydesmid millipede. *Ohio Journal of Science,* **27,** 25–43.

Miller, P. F. (1974). Competition between *Ophyiulus pilosus* (Newport) and *Iulus scandinavius* Latzel. *Symposia of the Zoological Society of London,* **32,** 553–74.

Minelli, A. (1990) (ed.). *Proceedings of the 7th International Congress of Myriapodology.* E. J. Brill, Leiden.

Minelli, A. and Bortoletto, S. (1988). Myriapod metamerism and arthropod segmentation. *Biological Journal of the Linnean Society,* **33,** 323–43.

Mitra, T. R. (1976). Millipedes entering houses. *Entomologists Monthly Magazine,* **112,** 44.

Mittelstaedt, M. L., Mittelstaedt, H., and Mohren, W. (1979). Interaction of gravity and idiothetic course control in millipedes. *Journal of Comprative Physiology,* **133A,** 267–81.

Moffett, D. F. (1975). Sodium and potassium transport across the isolated hindgut of the desert millipedes *Orthoporus ornatus. Comparative Biochemistry and Physiology,* **50A,** 57–63.

Morse, M. (1903). Unusual abundance of a myriapod, *Parajulus pennsylvanicus* (Brandt). *Science,* **18,** 59–60.

Mukhopadhyaya, M. C. and Saha, S. K. (1981). Observations on the natural population and sexual behaviour of *Orthomorpha coarctata* (Polydesmida, Paradoxosomatidae), a millipede of decaying wood and litters. *Pedobiologia,* **21,** 357–64.

Munoz-Cuevas, A. (1984). Photoreceptor structures and vision in arachnids and myriapods. In *Photoreception and vision in invertebrates,* (ed. M. A. Ali), pp. 335–99. Plenum Press, New York (NATO ASI Series, Series A, Life Sciences, Volume 74).

Murakami, Y. (1962). Post embryonic development of the common Myriapoda of Japan. XI. Life history of *Bazillozonium nodulosum* (Colobognatha, Platydesmidae). *Zoological Magazine* (*Tokyo*), **71,** 250–5.

Murakami, Y. (1963). Post embryonic development of the common Myriapoda of Japan. XIII. Life history of *Bazillozonium nodulosum* Verhoeff (Colobognatha, Platydesmidae). *Zoological Magazine* (*Tokyo*), **72,** 40–7.

Murakami, Y. (1965). Post embryonic development of the common Myriapoda of Japan. XVIII. Life history of *Ampelodesmus ivonis* MURAKAMI (Diplopoda, Cryptodesmidae). 2. *Zoological Magazine* (*Tokyo*), **74,** 31–7.

Nair, V. S. K. (1973). Neurosecretory flow in the perioesophageal tract of *Jonespeltis splendidus* (Diplopoda, Myriapoda). *Experentia,* **29,** 207–8.

Nair, V. S. K. (1980). Moult inhibition in the insect *Dysdercus cingulatus* (Insecta; Heteroptera) by the cerebral glands of the millipede *Jonespeltis splendidus* (Myriapoda: Diplopoda). *Experentia,* **38,** 607–9.

Needham, A. E. (1968). Integumental pigments of the millipede *Polydesmus angustus* (Latzel). *Nature,* **217,** 975–7.

Neuhauser, E. and Hartenstein, R. (1976). On the presence of O-dimethylase activity in invertebrates. *Comparative Biochemistry and Physiology,* **53C,** 37–9.

Neuhauser, E. and Hartenstein, R. (1978). Phenolic content and palatability of leaves and wood to soil isopods and diplopods. *Pedobiologia,* **18,** 99–109.

Neuhauser, E., Hartenstein, E. and Connors, W. J. (1978). Soil invertebrates and the degradation of vanillin, cinnamic acid, and lignins. *Soil Biology and Biochemistry,* **10,** 431–5.

Neumann, W. (1985). Veränderungen am mitteldarm von *Oxidus gracilis* (C. L. Koch, 1847) während einer häutung (Diplopoda). *Bijdragen tot de Dierkunde,* **55,** 149–58.

Newport, G. (1841). On the organs of reproduction, and the development of the Myriapoda. *Philosophical Transactions of the Royal Society of London,* **1841,** 99–130.

Newport, G. (1843). VIII. On the structure, relations and development of the nervous and circulatory systems and on the existence of a complete circulation of the blood in vessels in Myriapoda and macrourous Arachnida. *Philosophical Transactions of the Royal Society of London,* **1843,** 243–302.

Nguyen Duy-Jacquemin, M. (1972*a*). Régénération antennaire chez les larves et les adultes de *Polyxenus lagurus* (Diplopode, Pénicillate). *Compte Rendu Hebdomadaire des Séances de l'Académie des Sciences, Paris,* **274D,** 1323–6.

Nguyen Duy-Jacquemin, M. (1972*b*). Description d'un organe sensoriel antennaire cupuliforme chez *Polyxenus lagurus* (Diplopode, Pénicillates). *Compte Rendu Hebdomadaire des Séances de l'Académie des Sciences, Paris,* **275D,** 251–3.

Nguyen Duy-Jacquemin, M. (1974). Les organes intracérébraux de *Polyxenus lagurus* et comparison avec les organes neuraux d'autres diplopodes. *Symposia of the Zoological Society of London,* **32,** 211–16.

Nguyen Duy-Jacquemin, M. (1981). Ultrastructure des organes sensoriels de l'antenne de *Polyxenus lagurus* (Diplopode, Penicillate). I. Les cônes sensoriels apicaux du 8e article antennaire. *Annales des Sciences Naturelles, Zoologie,* Série 13, **3,** 95–114.

Nguyen Duy-Jacquemin, M. (1982). Ultrastructure des organes sensoriels de l'antenne de *Polyxenus lagurus* (Diplopode, Pénicillate). II. Les sensilles basiconiques des 6e et 7e articles antennaires. *Annales des Sciences Naturelles, Zoologie,* Série 13, **4,** 211–29.

Nguyen Duy-Jacquemin, M. (1983). Ultrastructure des organes sensoriels de l'antenne de *Polyxenus lagurus* (Diplopode, Pénicillate). III. Les sensilles coeloconiques des 6e et 7e articles antennaires. *Annales des Sciences Naturelles, Zoologie,* Série 13, **5,** 207–20.

Nguyen Duy-Jacquemin, M. (1985*a*). Ultrastructure des cônes sensoriels apicaux et des sensilles basiconiques spiniformes du 7e article des antennes de *Typhloblaniulus lorifer* et *Cylindroiulus punctatus* (Diplopodes, Iulides). *Annales des Sciences Naturelles, Zoologie,* Série 13, **7,** 67–88.

Nguyen Duy-Jacquemin, M. (1985*b*). Structures dendritiques des cônes antennaires apicaux de diplopodes (Myriapoda). *Bijdragen tot de Dierkunde,* **55,** 159–70.

Nguyen Duy-Jacquemin, M. (1988). Ultrastructure des organes sensoriels de l'antenne de *Polyxenus lagurus* (Diplopode, Pénicillate). IV. Les sensilles sétiformes à base renflée. *Annales des Sciences Naturelles, Zoologie,* Série 13, **9,** 161–75.

Nguyen Duy-Jacquemin, M. (1989). Ultrastructure des sensilles basiconiques bacilliformes des antennes du diplopode cavernicole *Typhloblaniulus lorifer* Bröl (Myriapode, Diplopode). *Memoires de Biospéologie,* **16,** 251–6.

Nguyen Duy-Jacquemin, M. (1990). Connaissances actuelles déduites de l'étude ultrastructurale des sensilles, sur le rôle de l'antenne dans la perception des stimuli chez les myriapodes. In *Proceedings of the 7th International Congress of Myriapodology,* (ed. A. Minelli), pp. 97–108, E. J. Brill, Leiden.

Nguyen Duy-Jacquemin, M. and Goyffon, M. (1977). Utilisation des protéinogrammes pour séparer les formes bisexuée et parthénogénétiques de *Polyxenus lagurus* (L.) (Diplopode, Pénicillate). *Compte Rendu Hebdomadaire des Séances de l'Académie des Sciences, Paris,* **284D,** 2047–9.

Nicholson, P. B., Bocock, K. L., and Heal, O. W. (1966). Studies on the decomposition of the faecal pellets of a millipede (*Glomeris marginata* (Villers)). *Journal of Ecology,* **54,** 755–66.

Nielsen, C. O. (1962). Carbohydrases in soil and litter invertebrates. *Oikos,* **13,** 200–15.

Niijima, K. and Shinohara, K. (1988). Outbreaks of the *Parafontaria laminata* group (Diplopoda: Xystodesmidae). *Japanese Journal of Ecology,* **38,** 257–68 (in Japanese).

Nunez, F. S. and Crawford, C. S. (1976). Digestive enzymes of the desert millipede *Orthoporus ornatus* (Girard) (Diplopoda: Spirostreptidae). *Comparative Biochemistry and Physiology,* **55A,** 141–5.

Nunez, F. S. and Crawford, C. S. (1977). Anatomy and histology of the alimentary tract of the desert millipede *Orthoporus ornatus* (Diplopoda: Spirostreptidae). *Journal of Morphology,* **151,** 121–30.

Ohba, M. and Aizawa, K. (1979). Multiplication of *Chilo* iridescent virus in non insect arthropods. *Journal of Invertebrate Pathology,* **33,** 278–83.

O'Neill, R. V. (1967). Niche segregation in seven species of Diplopods. *Ecology,* **48,** 983.

O'Neill, R. V. (1969). Adaptive responses to desiccation in the millipede *Narceus americanus* (Beauvois). *American Midland Naturalist,* **81,** 578–83.

O'Neill, R. V. and Reichle, D. E. (1970). Urban infestation by the millipede, *Oxidus gracilis. Journal of the Tennessee Academy of Science,* **45,** 114–15.

Oudejans, R. C. H. M. (1972*a*). Hydrocarbons in the millipede *Graphidostreptus tumuliporus* (Karsch) (Myriapoda, Diplopoda). I. *In vivo* incorporation of ^{14}C-labelled precursors into the hydrocarbon fraction. *Comparative Biochemistry and Physiology,* **42B,** 15–22.

Oudejans, R. C. H. M. (1972*b*). Composition of the saturated hydrocarbons from males, females and eggs of the millipede *Graphidostreptus tumuliporus. Journal of Insect Physiology,* **18,** 857–63.

Oudejans, R. C. H. M. and Horst, D. J. van der (1978). Cyclopropane fatty acids in millipedes: their occurrence and metabolism. *Abhandlungen und Verhandlungen des Naturwissenschaftlichen Vereins in Hamburg,* **21/22,** 345–8.

Oudejans, R. C. H. M. and Zandee, D. I. (1973). The biosynthesis of the hydrocarbons in males and females of the millipede *Graphidostreptus tumuliporus. Journal of Insect Physiology,* **19,** 2245–53.

Oudejans, R. C. H. M., Horst, D. J. van der, Opmeer, F. A., and Tieleman, W. J. (1976). On the function of cyclopropane fatty acids in millipedes (Diplopoda). *Comparative Biochemistry and Physiology,* **54B,** 227–30.

Owen, C. D. D. (1742). *An essay towards the natural history of serpents.* London.

Paoletti, M. G., Iovane, E., and Cortese, M. (1988). Pedofauna bioindicators and heavy metals in five agroecosystems in north-east Italy. *Revue d'Ecologie et de Biologie du Sol,* **25,** 33–58.

Pass, G. (1991). Antennal circulatory organs in Onychophora, Myriapoda and Hexapoda: functional morphology and evolutionary implications. *Zoomorphology,* **110,** 145–64.

Pedroli-Christen, A. (1978). Contribution à la connaissance du développement post-embryonnaire de *Craspedosoma alemannicum* (Verhoeff) et de *Xylophageuma zschokkei* Bigler (Diplopoda, Nematophora) dans une tourbière de Haut-Jura Suisse. *Revue Suisse de Zoologie,* **85,** 673–9.

Pedroli-Christen, A. (1990). Field investigations on *Rhymogona cervina* (Verhoeff) and *Rhymogona silvatica* (Rothenbuehler) (Diplopoda): Morphology, distribution and hybridization. In *Proceedings of the 7th International Congress of Myriapodology,* (ed. A. Minelli), pp. 27–43. E. J. Brill, Leiden.

Peitsalmi, M. (1974). Vertical orientation and aggregations of *Proteroiulus fuscus* (Am Stein) (Diplopoda, Blaniulidae). *Symposia of the Zoological Society of London,* **32,** 471–83.

Peitsalmi, M. (1981). Population structure and seasonal changes in activity of *Proteroiulus fuscus* (Am Stein) (Diplopoda, Blaniulidae). *Acta Zoologica Fennica,* **161,** 1–66.

Penteado, C. H. S. (1987). Respiratory responses of the tropical millipede, *Plusioporus setiger* (Broelemann, 1902) (Spirostreptida, Spirostreptidae) to normoxic and hypoxic conditions. *Comparative Biochemistry and Physiology,* **86A,** 163–8.

Penteado, C. H. S. and Hebling-Beraldo, M. J. A. (1991). Respiratory responses in a Brazilian millipede, *Pseudoannolene tricolor*, to declining oxygen pressures. *Physiological Zoology,* **64,** 232–41.

Penteado, C. H. S., Hebling-Beraldo, M. J. A., and Menes, E. G. (1991). Oxygen consumption related to size and sex in the tropical millipede *Pseudonannolene tricolor* (Diplopoda, Spirostreptida). *Comparative Biochemistry and Physiology,* **98A,** 265–9.

Percy, J. E. and Weatherstone, J. (1971). Studies of physiologically active arthropod secretions. V. Histological studies of the defence mechanism of *Narceus annularis* (Raf.) (Diplopoda: Spirobolida). *Canadian Journal of Zoology,* **49,** 278–9.

Peterson, S. C. (1986). Breakdown products of cyanogenesis repellency and toxicity to predatory ants. *Naturwissenschaften,* **73,** 627–8.

Petit, J. (1970). Sur la nature et l'accumulation de substances minérales dans les ovocytes de *Polydesmus complanatus* (Myriapode: Diplopode). *Compte Rendu Hebdomadaire des Séances de l'Académie des Sciences, Paris,* **270D,** 2107–10.

Petit, G. (1973). Étude morphologique et expérimentale de la métamorphose d'un appendice ambulatoire en gonopode chez le diplopode *Polydesmus angustus* Latz. *Annales d'Embryologie et de Morphogenèse,* **6,** 137–49.

Petit, G. (1974*a*). Sur les modalités de la croissance et la régénération des antennes de larves de *Polydesmus angustus* Latzel. *Symposia of the Zoological Society of London,* **32,** 301–15.

Petit, J. (1974*b*). Contribution à l'étude de l'appareil génital mâle et de la spermatogenèse chez *Polydesmus angustus* Latzel, Myriapode Diplopode. *Symposia of the Zoological Society of London,* **32,** 249–59.

Petit, J. (1976). Developpement comparé des appendices copulateurs (gonopodes) chez *Polydesmus angustus* Latzel et *Brachydesmus superus* Latzel (Diplopodes: Polydesmidae). *International Journal of Insect Morphology and Embryology,* **5,** 1261–72.

Petit, J. and Sahli, F. (1975). Cytochemical and electron-microscopy study of the paraoesophageal bodies and related nerves in *Schizophyllum sabulosum* (L.) Diplopoda Iulidae. *Cell and Tissue Research,* **162,** 367–75.

Petit, J. and Sahli, F. (1977). Étude ultrastructurale des cellules neurosécrétrices protocerebrales des globuli I chez *Schizophyllum sabulosum* (L.), (Diplopoda, Iulidae). *Bulletin de la Societe Zoologique de France,* **102,** 431–7.

Petit, J. and Sahli, F. (1978*a*). Étude ultrastructurale des plages paracommissurales (organes neurohémaux céphaliques) des Diplopodes Iulidae. *Abhandlungen und Verhandlungen des Naturwissenschaftlichen Vereins in Hamburg,* **21/22,** 295–309.

Petit, J. and Sahli, F. (1978*b*). Structure fine de quelques caractères sexuels secondaires chez les Diplopodes Iulides. *Abhandlungen und Verhandlungen des Naturwissenschaftlichen Vereins in Hamburg,* **21/22,** 183–95.

Pierrard, G. and Biernaux, J. (1974). Note à propos des diplopodes nuisibles aux cultures tempérées et tropicales. *Symposia of the Zoological Society of London,* **32,** 629–43.

Poinar, G. O. (1986). *Rhabditis myriophila* sp.n. (Rhabditidae: Rhabditida), associated with the millipede, *Oxidus gracilis* (Polydesmida: Diplopoda). *Proceedings of the Helminthological Society of Washington,* **53,** 232–6.

Poinar, G. O. and Thomas, G. M. (1985). Effect of neoaplectanid and heterorhabditid nematodes (Nematoda: Rhabditoidea) on the millipede *Oxidus gracilis*. *Journal of Invertebrate Pathology,* **45,** 231–5.

Ponomarenko, A. N., Trufanov, G. V., and Golubev, S. N. (1974). Microelements in soil invertebrates. *Soviet Journal of Ecology,* **5,** 279–81.

Prabhu, V. K. K. (1961). The structure of the cerebral glands and connective bodies of *Jonespeltis splendidus* Verhoeff (Myriapoda: Diplopoda). *Zeitschrift für Zellforschung und Mikroskopische Anatomie,* **54,** 717–33.

Price, P. W. (1988). An overview of organismal interactions in ecosystems in evolutionary and ecological time. *Agriculture, Ecosystems and Environment,* **24,** 369–77.

Pugach, S. and Crawford, C. S. (1978). Seasonal changes in hemolymph amino acids, proteins and inorganic ions of a desert millipede *Orthoporus ornatus* (Girard) (Diplopoda: Spirostreptidae). *Canadian Journal of Zoology,* **56,** 1460–5.

Radford, A. J. (1975). Millipede burns in man. *Tropical and Geographical Medicine,* **27,** 279–87.

Rajulu, S. G. (1967). Physiology of the heart of *Cingalobolus bugnioni (Diplopoda: Myriapoda). Experentia,* **23,** 388.

Rajulu, S. G. (1974). A comparative study of the organic components of the haemolymph of a millipede *Cingalobolus bugnioni* and a centipede *Scutigera longicornis* (Myriapoda). *Symposia of the Zoological Society of London,* **32,** 347–64.

Ramsey, J. M. (1966). Vast migrating armies of the millipede *Pseudopolydesmus serratus* (Say) in the Dayton region. *Ohio Journal of Science,* **66,** 339.

Rantala, M. (1974). Sex ratio and periodomorphosis of *Proteroiulus fuscus* (Am Stein) (Diplopoda, Blaniulidae). *Symposia of the Zoological Society of London,* **32,** 463–9.

Rantala, M. (1990). Intensity of radiation in some diplopods and chilopods reared in radioactive compost. In *Proceedings of the 7th International Congress of Myriapodology,* (ed. A. Minelli), pp. 271–87. E. J. Brill, Leiden.

Ravindranath, M. H. (1973). The hemocytes of a millipede *Thyropygus poseidon. Journal of Morphology,* **141,** 257–67.

Ravindranath, M. H. (1974). Changes in the population of circulating hemocytes during molt cycle phases of the millipede *Thyropygus poseidon. Physiological Zoology,* **47,** 252–60.

Ravindranath, M. H. (1981). Onychophorans and Myriapods. In *Invertebrate blood cells,* Volume 2, (ed. N. A. Ratcliffe and A. F. Rowley), pp. 273–54. Academic Press, London.

Read, H. J. (1985). Stadial distributions of *Ommatoiulus moreleti* at different altitudes in Madeira with reference to life history phenomena (Diplopoda; Julidae). *Bijdragen tot de Dierkunde,* **55,** 177–80.

Read, H. J. (1988). The life histories of millipedes: a review of those found in British species of the order Julida and comments on endemic Madeiran *Cylindroiulus* species. *Revue d'Ecologie et de Biologie du Sol,* **25,** 451–67.

Read, H. J. (1989). New species and records of the *Cylindroiulus madeirae*-group with notes on phylogenetic relationships (Diplopoda, Julida: Julidae). *Entomologica Scandinavica,* **19,** 333–47.

Read, H. J. (1990). The generic composition and relationships of the Cylindroiulini—a cladistic analysis (Diplopoda, Julida: Julidae). *Entomologica Scandinavica,* **21,** 97–112.

Read, H. J. and Martin, M. H. (1990). A study of millipede communities in woodlands contaminated with heavy metals. In *Proceedings of the 7th International Congress of Myriapodology,* (ed. A. Minelli), pp. 289–98. E. J. Brill, Leiden.

Reger, J. F. (1971). Studies on the fine structure of spermatids and spermatozoa from the millipede *Spirobolus* sp. *Journal of Submicroscopic Cytology,* **3,** 33–44.

Reger, J. F. and Cooper, D. P. (1968). Studies on the fine structure of spermatids and spermatozoa from the millipede *Polydesmus* sp. *Journal of Ultrastructural Research,* **23,** 60–70.

Reger, J. F. and Fitzgerald, M. E. (1979). The fine structure of membrane complexes in spermatozoa of the millipede *Spirobolus* sp., as seen by thin section and freeze-fracture techniques. *Journal of Ultrastructural Research,* **67,** 95–108.

Reichle, D. E. (1968). Relation of body size to food intake, oxygen consumption and trace element metabolism in forest floor arthropods. *Ecology,* **49,** 538–42.

Reichle, D. E. (1969). Measurement of elemental assimilation by animals from radioisotope retention patterns. *Ecology,* **50,** 1102–4.

Remy, P. (1950). On the enemies of Myriapods. *Naturalist,* **1950,** 103–8.

Rettenmeyer, C. W. (1962). The behaviour of millipedes found with neotropical army ants. *Journal of the Kansas Entomological Society,* **35,** 377–84.

Riddle, W. A. (1985). Hemolymph osmoregulation in several myriapods and arachnids. *Comparative Biochemistry and Physiology,* **80A,** 313–23.

Riddle, W. A., Crawford, C. S., and Zeitone, A. M. (1976). Patterns of haemolymph osmoregulation in three desert arthropods. *Journal of Comparative Physiology,* **112B,** 295–305.

Rolfe, W. D. I. (1986). Aspects of the Carboniferous terrestrial arthropod community. *Ninth International Congress on Carboniferous Stratigraphy and Geology,* IX–ICC, **5,** 303–16. Washington DC and Urbana Illinois.

Rolfe, W. D. I. and Ingham, J. K. (1967). Limb structure, affinity and diet of the Carboniferous 'centipede' *Arthropleura. Scottish Journal of Geology,* **3,** 118–24.

Roncadori, R. W., Duffey, S. S., and Blum, M. S. (1985). Antifungal activity of defensive secretions of certain millipedes. *Mycologia,* **77,** 185–91.

Röper, H. and Heyns, K. (1977). Spurenanalytik von p-Benzochinon und Hydrochinon—Derivaten mit Gaschromatographie und Gaschromatographie/ Massenspektrometrie. Identifizierung von Wehrsekret-Komponenten europäischer Juliden. *Zeitschrift für Naturforschung,* **32C,** 61–6.

Rosenberg, M. and Warburg, M. R. (1982). The brain and neurosecretory cells of a desert millipede and their relation to the reproductive cycle. *General and Comparative Endocrinology,* **46,** 371.

Rossi, W. and Balazuc, J. (1977). Laboulbéniales parasites de Myriapodes. *Revue de Mycologie,* **41,** 525–35.

Roth-Holzapfel, M. (1990). Multi-element analysis of invertebrate animals in a forest ecosystem (*Picea abies* L.). In *Element concentrations cadasters in ecosystems,* (ed. H. Leith and B. Markert), pp. 192–8. VCH, Weinheim.

Rust, M. K. and Reierson, D. A. (1977). Effectiveness of barrier toxicants against migrating millipedes. *Journal of Economic Entomology,* **70,** 477–9.

Sadek, R. A. (1981). The diet of the Madeiran lizard *Lacerta dugesii. Zoological Journal of the Linnean Society of London,* **73,** 313–41.

Sahli, F. (1958*a*). Quelques données sur la neurosécrétion chez le Diplopode *Tachypodoiulus albipes* C. L. Koch. *Compte Rendu Hebdomadaire des Séances de l'Académie des Sciences, Paris,* **246,** 470–2.

Sahli, F. (1958*b*). Donnees sur le développement post-embryonnaire du Diplopode *Tachypodoiulus albipes* C. L. Koch. *Compte Rendu Hebdomadaire des Séances de l'Académie des Sciences, Paris,* **246,** 2037–9.

Sahli, F. (1961*a*). Sur une formation hypocérébrale chez les Diplopodes Julides. *Compte Rendu Hebdomadaire des Séances de l'Académie des Sciences, Paris,* **252,** 2443–4.

Sahli, F. (1961*b*). La succession des differentes formes males au cours de la periodomorphose chez le Diplopode *Tachypodoiulus albipes* C. L. Koch. *Compte Rendu Hebdomadaire des Séances de l'Académie des Sciences, Paris,* **253,** 3094–5.

Sahli. F. (1962). Sur le système neurosécréteur du Polydesmoide *Orthomorpha gracilis* C. L. Koch (Myriapoda, Diplopoda). *Compte Rendu Hebdomadaire des Séances de l'Académie des Sciences, Paris,* **254D,** 1498–500.

Sahli, F. (1968). Sur l'existence de mâles intercalaires chez le diplopode *Cylindroiulus nitidus* (Verhoeff 1891). *Compte Rendu Hebdomadaire des Séances de l'Académie des Sciences, Paris,* **266D,** 360–3.

Sahli, F. (1972). Modifications des caractères sexuels secondaires mâles chez les Iulidae (Myriapoda, Diplopoda) sous l'influence de Gordiacés parasites. *Compte Rendu Hebdomadaire des Séances de l'Académie des Sciences, Paris*, **274D,** 900–3.

Sahli, F. (1974). Sur les organes neurohémaux et endocrines des Myriapodes Diplopodes. *Symposia of the Zoological Society of London,* **32,** 217–30.

Sahli, F. (1977a). Présence de corps para-oesophagiens chex les Nematophora Lysiopetalidae (Myriapoda, Diplopoda). *Compte Rendu Hebdomadaire des Séances de l'Académie des Sciences, Paris,* **284D,** 211–13.

Sahli, F. (1977*b*). Sur les cellules neurosécrétrices des globuli I et sur la voie neurosécrétrice protocéphaliques des Myriapodes Diplopodes. Premières observations chez les Polydesmida Platyrrhacidae, les Nematophora Lysiopetalidae, les Spirostreptida et les Spirobolida. *Compte Rendu Hebdomadaire des Séances de l'Académie des Sciences, Paris,* **284D,** 815–17.

Sahli, F. (1979). Different types of neurosecretory system in Diplopoda. In *Myriapod biology,* (ed. M. Camatini), pp. 279–85. Academic Press, London.

Sahli, F. (1985*a*). Neurohaemal organs in Myriapoda and phylogeny. *Bijdragen tot de Dierkunde,* **55,** 193–201.

Sahli, F. (1985*b*). Periodomorphose chez *Cylindroiulus nitidus* (Verhoeff) en Grand-Bretagne et en Allemagne (Diplopoda: Julidae). *Bijdragen tot de Dierkunde,* **55,** 190–2.

Sahli, F. (1986). On some roles of periodomorphosis in *Ommatoiulus sabulosus* (L.) (Myriapoda, Diplopoda) in the maritime Alps. *Advances in Invertebrate Reproduction,* **4,** 409–16.

Sahli, F. (1990). On post-adult moults in Julida (Myriapoda, Diplopoda). Why do periodomorphosis and intercalaries occur in males? In *Proceedings of the 7th International Congress of Myriapodology,* (ed. A. Minelli), pp. 135–56, E. J. Brill, Leiden.

Sahli, F. and Petit, J. (1971). Observations sur l'ultrastructure des organes de Gabe des Polydesmidae et des Iulidae (Diplopoda). *Compte Rendu Hebdomadaire des Séances de l'Académie des Sciences, Paris,* **275D,** 2017–20.

Sahli, F. and Petit, J. (1973). Observations sur l'ultrastructure des corps paraoesophagiens des Diplopodes Iulides. *Compte Rendu Hebdomadaire des Séances de l'Académie des Sciences, Paris,* **276D,** 2019–22.

Sahli. F. and Petit, J. (1974). Observations sur l'ultrastructure des corps connectifs (organes neurohémaux céphaliques) d'*Orthomorpha gracilis* (C. L. K.) (Diplopoda, Polydesmoidea). *Compte Rendu Hebdomadaire des Séances de l'Académie des Sciences, Paris,* **279D,** 2055–8.

Sahli. F. and Petit, J. (1975). Les plages paracommissurales (formations neurohémales cephaliques) des Diplopodes. *Compte Rendu Hebdomadaire des Séances de l'Académie des Sciences, Paris,* **280D,** 2001–4.

Sahli, F. and Petit, J. (1979). The latero-oesophageal complex in Julidae (Diplopoda). In *Myriapod biology,* (ed. M. Camatini), pp. 307–13. Academic Press, London.

Saita, A. and Candia-Carnevali, M. D. (1978). Neuromuscular junctions in Myriapoda: electron microscopic observations. *Abhandlungen und Verhandlungen des Naturwissenschaftlichen Vereins in Hamburg,* **21/22,** 279–93.

Saki, E. (1934). Diplopoda obstructive to railway traffic. *Botany and Zoology* (Tokyo), **2,** 821–33.

Sakwa, W. N. (1974). A consideration of the chemical basis of food preference in millipedes. *Symposia of the Zoological Society of London,* **32,** 329–46.

Sakwa, W. N. (1978). An electrophysiological analysis of receptor cell activity in millipede chemoreceptor sensilla. *Abhandlungen und Verhandlungen des Naturwissenschaftlichen Vereins in Hamburg,* **21/22,** 311–20.

Satyam, P. and Ramamurthi, R. (1978). Neuroendocrine control of metabolism in millipedes: *Spirostreptus asthenes. Abhandlungen und Verhandlungen des Naturwissenschaftlichen Vereins in Hamburg,* **21/22,** 325–7.

Saudray, Y. (1953). Développement post-embryonnaire d'un Iulide indigene *Cylindroiulus (Aneuloboiulus) silvarum* Meinert. *Archives de Zoologie Expérimentale et Générale,* **89,** 1–14.

Saulnier, L. and Athias-Binche, F. (1986). Modalités de la cicatrisation des écosystemes méditerranéens après incendie: cas de certains Arthropodes du sol. 2. Les Myriapodes édaphiques. *Vie et Milieu,* **36C,** 191–204.

Schildknecht, H. and Wenneis, W. F. (1967). Über Arthropoden-Abwehrstoffe. XXV. Anthranilsäure als precursor der Arthropoden-Alkaloide Glomerin und Homoglomerin. *Tetrahedron Letters,* **19,** 1815–18.

Schildknecht, H., Wenneis, W. F., Weis, K. H., and Maschwitz, U. (1966). Glomerin, ein neues Arthropoden-Alkaloid. *Zeitschrift für Naturforschung,* **21B,** 121–7.

Schlüter, U. (1979). The ultrastructure of an exocrine gland complex in the hindgut of *Scaphiostreptus* sp. (Diplopoda: Spirostreptidae). In *Myriapod biology,* (ed. M. Camatini), pp. 143–55. Academic Press, London.

Schlüter, U. (1980*a*). Ultrastruktur der Pyloruszähnchen zweier Tausendfüssler (*Tachypodoiulus niger, Polydesmus angustus*). *Acta Zoologica* (Stockholm), **61,** 171–8.

Schlüter, U. (1980*b*). Die Feinstruktur der pylorusdrüsen von *Polydesmus angustus* Latzel und *Glomeris marginata* Villers (Diplopoda). *Zoomorphology,* **94,** 307–19.

Schlüter, U. (1980*c*). Plasmalemma–mitochondrial complexes involved in water transport in the hindgut of a millipede *Scaphiostreptus* sp. *Cell and Tissue Research,* **205,** 333–6.

Schlüter, U. (1980*d*). Cytopathological alterations in the hindgut of a millipede induced by atypical diet. *Journal of Invertebrate Pathology,* **36,** 133–5.

Schlüter, U. (1982). The anal glands of *Rhapidostreptus virgator* (Diplopoda, Spirostreptidae). I. Appearance during the intermoult cycle. *Zoomorphology,* **100,** 65–73.

Schlüter, U. (1983). The anal glands of *Rhapidostreptus virgator* (Diplopoda, Spirostreptidae). II. Appearance during a moult. *Zoomorphology,* **102,** 79–86.

Schlüter, U. and Seifert, G. (1985*a*). Functional morphology of the hindgut–malpighian tubule-complex in *Polyxenus lagurus* (Diplopoda; Penicillata). *Bijdragen tot de Dierkunde,* **55,** 209–18.

Schlüter, U. and Seifert, G. (1985*b*). Rickettsiales-like microorganisms associated with the Malpighian tubules of the millipede *Polyxenus lagurus* (Diplopoda; Penicillata). *Journal of Invertebrate Pathology,* **46,** 211–14.

Schmidt, P. (1985). Beiträge zur kenntnis der Niederen Myriapoden. *Zeitschrift für Wissenschaftliche Zoologie,* **59,** 436–510.

Schömann, K. (1956). Zur Biologie von *Polyxenus lagurus* (L. 1758). *Zoologische Jahrbücher (Systematik),* **84,** 195–256.

Schubart, O. (1934). Diplopoda. *Tierwelt Deutschlands,* **28,** 1–318.

Schulte, F. (1989*a*). The association between *Rhabditis necromena* Sudhaus & Schulte 1989 (Nematoda: Rhabditidae) and native and introduced millipedes in south Australia. *Nematologica,* **35,** 82–9.

Schulte, F. (1989*b*). Biologische Kontrolle eingeschleppter Tausendfüsser in Sud-Australien. *Biologie in Unserer Zeit,* **6,** 203–4.

Scott, H. (1958*a*). Migrant millipedes and centipedes entering houses 1953–1957. *Entomologists Monthly Magazine,* **94,** 73–7.

Scott, H. (1958*b*). Migrant millipedes entering houses 1958. *Entomologists Monthly Magazine,* **94,** 252–6.

Seastedt, T. R. and Tate, C. M. (1981). Decomposition rates and nutrient contents of arthropod remains in forest litter. *Ecology,* **62,** 13–19.

Seifert, B. (1932). Anatomie und biologie des Diplopoden *Strongylosoma pallipes. Zeitschrift für Morphologie und Ökologie der Tiere,* **25,** 362–507.

Seifert, G. (1971). Ein bisher unbekanntes Neurohämalorgan von *Craspedosoma rawlinsii* Leach (Diplopoda, Nematophora). *(Diplopoda, Nematophora). Zeitschrift fur Morphologie und Okologie der Tiere,* **70,** 128–40.

Seifert, G. (1979). Considerations about the evolution of excretory organs in terrestrial arthropods. In *Myriapod biology* (ed. M. Camatini), pp. 353–72. Academic Press, London.

Seifert, G. and Rosenberg, J. (1976). Die Ultrastruktur der Nephrozyten von *Orthomorpha gracilis* (C. L. Koch) (Diplopoda, Strongylosomidae). *Zoomorphology,* **85,** 23–7.

Seifert, G. and Rosenberg, J. (1977). Feinstruktur der leberzellen von *Oxidus gracilis* (C. L. Koch, 1847) (Diplopoda, Paradoxosomatidae). *Zoomorphology,* **88,** 145–62.

Shaw, G. G. (1966). New observations on reproductive behaviour in the millipede *Narceus annularis* (Raf.). *Ecology,* **47,** 322–3.

Shear, W. A. (1969). A synopsis of the cave millipedes of the United States with an illustrated key to genera. *Psyche,* **76,** 126–43.

Shear, W. A. (1981). Two fossil millipeds from the Dominican amber (Diplopoda: Chytodesmidae, Siphonophoridae). *Myriapodologica,* **1,** 51–4.

Shear, W. A. (1984). Cave millipedes of the United States. III. Two new species from the western states (Diplopoda: Polydesmida, Chordeumatida). *Myriapodologica,* **1,** 95–104.

Shear, W. A. (1991). The early development of terrestrial ecosystems. *Nature,* **351,** 283–9.

Shear, W. A. and Kukalova-Peck, J. (1990). The ecology of Paleozoic terrestrial arthropods: the fossil evidence. *Canadian Journal of Zoology,* **68,** 1807–34.

Shelley, R. M. (1976). Millipedes of the *Sigmoria latior* complex (Polydesmida: Xystodesmidae). *Proceedings of the Biological Society of Washington,* **89,** 17–37.

Shelley, R. M. (1977). Appendicular abnormalities in the milliped family Xystodesmidae (Polydesmida). *Canadian Journal of Zoology,* **55,** 1014–18.

Shelley, R. M. (1990). Are allopatric/parapatric mosaic complexes widespread in the Diplopoda? In *Proceedings of the 7th International Congress of Myriapodology,* (ed. A. Minelli), p. 23. E. J. Brill, Leiden.

Shinohara, K. (1981). On sprawling of myriapod injuries on city. *Japanese Journal of Sanitary Zoology,* **32,** 249–50.

Siegel, S. M., Siegel, B. Z., Puerner, N., Speitel, T., and Thorarinsson, F. (1975). Water and soil biotic relations in mercury distribution. *Water, Air and Soil Pollution,* **4,** 9–18.

Silvestri, P. (1903). *Diplopoda.* Portico.

Simonsen, Å. (1983). Cluster analysis of millipede communities of different altitudes and distances from the coast in Setesdalen, southern Norway. *Fauna Norvegica,* **31B,** 96–102.

Simonsen, Å. (1990). Phylogeny and biogeography of the millipede order Polydesmida with special emphasis on the suborder Polydesmida. Unpublished Ph.D. Thesis, University of Bergen.

Simkiss, K. (1976). Intracellular and extracellular routes in bio-mineralization. *Symposia of the Society for Experimental Biology,* **30,** 423–4.

Sinclair, F. G. (1895). Myriapoda. In *The Cambridge natural history,* Volume 5, (ed. S. F. Harmer and A. E. Shipley), pp. 29–80. Cambridge University Press.

Smith, K. G. V. (1973). *Insects and other arthropods of medical importance.* British Museum (Natural History), London.

Smolanoff, J., Kluge, A. F., Meinwald, J., McPhail, A., Miller, R. W., Hicks, K., and Eisner, T. (1975). Polyzonimine: a novel terpenoid insect repellent produced by a millipede. *Science,* **188,** 734–6.

Snider, R. M. (1981*a*). Growth and survival of *Polydesmus inconstans* (Diplopoda: Polydesmidae) at constant temperatures. *Pedobiologia,* **22,** 345–53.

Snider, R. M. (1981*b*). The reproductive biology of *Polydesmus inconstans* (Diplopoda: Polydesmidae) at constant temperatures. *Pedobiologia,* **22,** 354–65.

Snider, R. (1984*a*). Diplopoda as food for Coleoptera. Laboratory experiments. *Pedobiologia,* **26,** 197–204.

Snider, R. (1984*b*). The ecology of *Polydesmus inconstans* (Diplopoda: Polydesmidae) in Michigan wood lots. *Pedobiologia,* **26,** 185–96.

Spaull, V. W. (1976). The life-history and post-embryonic development of *'Spirobolus' bivirgatus* (Diplopoda: Spirobolida) on Aldabra, Western Indian Ocean. *Journal of Zoology,* **180,** 391–405.

Spies, T. (1981). Structure and phylogenetic interpretation of diplopod eyes (Diplopoda). *Zoomorphology,* **98,** 241–60.

Srivastava, P. D. and Srivastava, Y. N. (1967). *Orthomorpha* sp., a new predatory millipede on *Achatenia fulica* in Andamans. *Experentia,* **23,** 776.

Stachurski, A. and Zimka, J. (1968). Food preferences of frogs and the sex ratio in populations of some saprophages (Diplopoda, Isopoda). *Bulletin de l'Académie Polonaise des Sciences, Série des Sciences Biologiques,* **16,** 101–5.

Stamou, G. P. and Iatrou, G. D. (1990). Respiration metabolism of *Glomeris balcanica* at a constant temperature. In *Proceedings of the 7th International Congress of Myriapodology,* (ed. A. Minelli), pp. 197–205. E. J. Brill, Leiden.

Stebbins, R. C. (1944). Lizards killed by a millipede. *American Midland Naturalist,* **32,** 777–8.

Stephenson, J. W. (1961). The biology of *Brachydesmus superus* (Latzel). *Annals and Magazine of Natural History, Series 13,* **3,** 311–19.

Stewart, T. C. and Woodring, J. P. (1973). Anatomical and physiological studies of water balance in the millipedes *Pachydesmus crassicutis* (Polydesmida) and *Orthoporus texicolens* (Spirobolida). *Comparative Biochemistry and Physiology,* **44A,** 735–50.

Størmer, L. (1976). Arthropods from the Lower Devonian (Lower Emsian) of Alken an der Mosel, Germany. Part 5. Myriapoda and additional forms, with general remarks on fauna and problems regarding invasion of land by arthropods. *Senckenbergiana Lethaea,* **57,** 87–183.

Striganova, B. R. (1970). Cellulose decomposition in the intestine of the millipede *Pachyiulus foetidissimus* (Mur.) (Juloidea, Diplopoda). *Dokladȳ Akademii Nauk Soyuza Sovetskikh Sotsialisticheskikh Respublik,* **190,** 703–5.

Striganova, B. R. (1971). A comparative account of the activity of different groups of soil invertebrates in the decomposition of forest litter. *Soviet Journal of Ecology,* **2,** 316–21.

Striganova, B. R. and Prishutova, Z. G. (1990). Food requirements of diplopods in the dry steppe subzone of the USSR. *Pedobiologia,* **34,** 37–41.

Striganova, B. R. and Rachmanov, R. R. (1972). Comparative study of the feeding activity of Diplopoda in Lenkoran province of Azerbaijan. *Pedobiologia,* **12,** 430–3.

Subramoniam, T. (1971). Peroxidase uptake by the fat body of a millipede *Spirostreptus asthenes* (Diplopoda: Myriapoda). *Experentia,* **27,** 1296–7.

Subramoniam, T. (1972). Studies on the fat body of millipedes. I. Histological and histochemical features of the fat body of the millipede *Spirostreptus asthenes* Pocock. *Zoologischer Anzeiger,* **189,** 200–8.

Subramoniam, T. (1974). A histochemical study on the cuticle of a millipede *Spirostreptus asthenes* (Diplopoda: Myriapoda). *Acta Histochemica,* **51,** 200–4.

Sutcliffe, D. W. (1963). The chemical composition of haemolymph in insects and some other arthropods, in relation to their phylogeny. *Comparative Biochemistry and Physiology,* **9,** 121–35.

Swift, M. J., Heal, O. W., and Anderson, J. M. (1979). *Decomposition in terrestrial ecosystems.* Blackwell Scientific, Oxford.

Tabacaru, I. (1969). Sur l'origine de la faune des diplopodes des Carpates. *Bulletin du Muséum National d'Histoire Naturelle, Série II,* **41,** 139–43.

Taylor, E. C. (1982*a*). Role of aerobic microbial populations in cellulose digestion by desert millipedes. *Applied Environmental Microbiology,* **44,** 281–91.

Taylor, E. C. (1982*b*). Fungal preference by a desert millipede *Orthoporus ornatus* (Spirostreptidae). *Pedobiologia,* **23,** 331–6.

Taylor, M. G. and Simkiss, K. (1984). Inorganic deposits in invertebrate tissues. *Environmental Chemistry,* **3,** 102–38.

Telford, S. R. and Dangerfield, J. M. (1990). Manipulation of the sex ratio and duration of copulation in the tropical millipede *Alloporus uncinatus*: a test of the copulatory guarding hypothesis. *Animal Behaviour,* **40,** 984–6.

Tichy, H. (1975). Unusual fine structure of sensory hair triad of the millipede *Polyxenus. Cell and Tissue Research,* **156,** 229–38.

Tiegs, O. W. (1947). The development and affinities of the Pauropoda based on a study of *Pauropus silvaticus*. *Quarterly Journal of Microscopical Science,* **88,** 165–267, 275–336.

Tinliang, J., Guiwen, F., Jianhua, S., Lanfang, L., and Xiangqi, F. (1981). Observation of the effect of *Spirobolus bungii* extract on cancer cells. *Journal of Traditional Chinese Medicine,* **1,** 34–8.

Tömösváry, E. (1883). Eigentümliche Sinnes Organe der Myriapoden. *Mathematische und Naturwissenschaftliche Berichte aus Ungarn,* **1,** 324–5.

Towers, G. H. N., Duffey, S. S., and Siegel, S. M. (1972). Defensive secretion: Biosynthesis of hydrogen cyanide and benzaldehyde from phenylalanine by a millipede. *Canadian Journal of Zoology,* **50,** 1047–50.

Toye, S. A. (1966*a*). The reactions of three species of Nigerian millipedes (*Spirostreptus assiniensis, Oxydesmus* sp., and *Habrodesmus falx*) to light, humidity and temperature. *Entomologia Experimentalis et Applicata,* **9,** 468–83.

Toye, S. A. (1966*b*). Studies on the locomotory activity of three species of Nigerian millipedes. *Entomologia Experimentalis et Applicata,* **9,** 369–77.

Toye, S. A. (1966*c*). The effect of desiccation on the behaviour of three species of Nigerian millipedes: *Spirostreptus assiniensis, Oxydesmus* sp., and *Habrodesmus falx. Entomologia Experimentalis et Applicata,* **9,** 378–84.

Toye, S. A. (1967). Observations on the biology of three species of Nigerian millipedes. *Journal of Zoology,* **152,** 67–78.

Tracz, H. (1987). The role of *Proteroiulus fuscus* (Am Stein 1857) (Diplopoda) in the circulation of some elements in the fresh pine forest. *Annals of Warsaw Agricultural University, Forestry and Wood Technology,* **36,** 101–7.

Upton, S. J., Crawford, C. S., and Hoffman, R. L. (1983). A new species of thelastomatid (Nematoda: Thelastomatidae) from the desert millipede *Orthoporus ornatus* (Diplopoda: Spirostreptidae). *Proceedings of the Helminthological Society of Washington,* **50,** 69–82.

Vachon, M. (1947). Contribution à l'étude de développement post-embryonnaire de *Pachybolus ligulatus* Voges. Les étapes de la croissance. *Annales des Sciences Naturelles, Zoologie,* (11), **9,** 109–21.

Vala, J. C., Bailey, P. T., and Gasc, C. (1980). Immature stages of the fly *Pelidnoptera nigripennis* (Fabricus) (Diptera: Phaeomyiidae), a parasitoid of millipedes. *Systematic Zoology,* **15,** 391–9.

Van Der Drift, J. (1951). Analysis of the animal community in a beech forest floor. *Tijdschrift voor Entomologie,* **94,** 1–168.

Van der Drift, J. (1975). The significance of the millipede *Glomeris marginata* (Villers) for oak-litter decomposition and an approach of its part in energy flow. In *Progress in soil zoology,* (ed. J. Vanek), pp. 293–8. Academia, Prague.

Van der Walt, A., McClain, E., Puren, A., and Savege, N. (1990). Phylogeny of arthropod immunity. An inducible humoral response in the Kalahari millipede, *Triaenostreptus triodus* (Attems). *Naturwissenschaften,* **77,** 189–90.

Verhoeff, K. W. (1900). Wandern de Doppelfülser, Eisenbahn zuge hemmend. *Zoologischer Anzeiger,* **23,** 465–73.

Verhoeff, K. W. (1923). Périodomorphose. *Zoologischer Anzeiger,* **56,** 233–8; 241–54.

Verhoeff, K. W. (1926–32). Diplopoda. *Bronn's Klassen und Ordnungen des Tierreichs,* **5,** (II, 1–2), 1–2084.

Voigtländer, K. (1987). Untersuchungen zur Bionomie von *Enantiulus nanus* (Latzel, 1884) und *Allajulus occultus* C. L. Kock 1847 (Diplopda, Julidae).

Abhandlungen und Berichte des Naturkundemuseums Forschungsstelle—Görlitz, **60,** 1–116.

Warburg, M. R. and Rosenberg, M. (1983). Cerebral neurosecretory cells in the millipede *Archispirostreptus syriacus* De Saussure (Diplopoda, Spirostreptidae). *Acta Zoologica* (Stockholm), **64,** 107–15.

Walker, L. J. and Crawford, C. S. (1980). Integumental ultrastructure of the desert millipede *Orthoporus ornatus* (Girard) (Diplopoda: Spirostreptidae). *Journal of Insect Morphology and Embryology,* **9,** 231–49.

Weatherston, J. and Percy, J. E. (1969). Studies of physiologically active arthropod secretions. III. Chemical, morphological and histological studies of the defence mechanism of *Uroblaniulus canadensis* (Say) (Diplopoda: Julida). *Canadian Journal of Zoology,* **47,** 1389–94.

Wegensteiner, R. (1982). Light microscopic and electro microscopic investigations on the cuticular structure of *Polyzonium germanicum* (Colobognatha, Diplopoda). *Mikroskopie,* **39,** 198–206.

Wernitzsch, W. (1910). Beiträge zur Kenntnis von *Craspedosoma simile* und des Trachensystems der Diplopoden. *Jenaische Zeitschrift für Naturwissenschaft,* **46,** 225–84.

West, W. R. (1953). An anatomical study of the male reproductive system of a Virginia millipede. *Journal of Morphology,* **93,** 123–76.

Wheeler, W. M. (1890). Hydrocyanic acid secreted by *Polydesmus virginiensis,* Drury. *Psyche,* **4,** 442.

Wheeler, J. M., Meinwald, J., Hurst, J. J., and Eisner, T. (1964). Trans 2-dodecenal and 2-methyl-1, 4-quinone produced by a millipede. *Science,* **144,** 540–1.

White, M. J. D. (1979). The present status of myriapod cytogenetics. In *Myriapod biology*, (ed. M. Camatini), pp. 3–8. Academic Press, London.

Wiley, E. O. (1981). *Phylogenetics. The Theory and Practice of Phylogenetic Systematics.* Wiley, New York.

Willey, R. B. and Brown, W. L. (1983). New species of the ant genus *Myopias* (Hymenoptera: Formicidae: Ponerinae). *Psyche,* **90,** 249–85.

Wilson, G. G. and Burke, J. M. (1972). A Rickettsia-like organism associated with the millipede *Ophyiulus pilosus* (Newport) (Diplopoda: Iulidae). *Canadian Journal of Microbiology,* **18,** 538.

Witz, B. W. (1990). Antipredator mechanisms in arthropods: a twenty year literature survey. *Florida Entomologist,* **73,** 71–99.

Wood, W. F., Shepherd, J., Chong, B., and Meinwald, J. (1975). Ubiquinone-O in the defensive spray of African millipede. *Nature,* **253,** 625–6.

Woodring, J. P. (1974). Effects of rapid and slow dehydration on the hemolymph osmolarity and Na^{+}–K^{+} concentration in the millipede *Pachydesmus crassicutis. Comparative Biochemistry and Physiology,* **49A,** 115–19.

Woodring, J. P. and Blum, M. S. (1965). The anatomy, physiology and comparative aspects of the repugnatorial glands of *Orthocricus arboreus* (Diplopoda: Spirobolida). *Journal of Morphology,* **116,** 99–108.

Wooten, R. C. and Crawford, C. S. (1974). Respiratory metabolism in the desert millipede *Orthoporus ornatus* (Girard) (Diplopoda). *Oecologia,* **17,** 179–86.

Wooten, R. C. and Crawford, C. S. (1975). Food, ingestion rates and assimilation in the desert millipede *Orthoporus ornatus* (Girard) (Diplopoda). *Oecologia,* **20,** 231–6.

Wooten, R. C., Crawford, C. S., and Riddle, W. A. (1975). Behavioural thermoregulation of *Orthoporus ornatus* (Diplopoda: Spirostreptidae) in three desert habitats. *Zoological Journal of the Linnean Society of London,* **57,** 59–74.

Wright, K. A. (1979). Trichomyetes and oxyuroid nematodes in the millipede *Narceus annularis. Proceedings of the Helminthological Society of Washington,* **46,** 213–23.

Wright, J. C. and Machin, J. (1990). Water vapour absorption in terrestrial isopods. *Journal of Experimental Biology,* **154,** 13–30.

Xylander, W. E. R. (1991). Immune defense reactions of Myriapoda. In *Proceedings of the 8th International Congress of Myriapodology. Veröffentlichungen der Universität Innsbruck.* (In press.)

Xylander, W. E. R. and Nevermann, L. (1990). Antibacterial activity in the haemolymph of Myriapods (Arthropoda). *Journal of Invertebrate Pathology,* **56,** 206–14.

Zacharuck, R. Y. (1985). Antennae and sensilla. In *Comprehensive insect physiology and biochemistry,* (ed. G. A. Kerkut and L. I. Gilbert). Volume 6, Nervous system: sensory, pp. 1–69. Pergamon Press, Oxford.

Zandee, D. I. (1967). Absence of cholesterol synthesis as contrasted with the presence of fatty acid synthesis in some arthropods. *Comparative Biochemistry and Physiology,* **20,** 811–22.

Zhulidov, A. V. and Sizova, M. G. (1985). Accumulation of lead from food by the millipede *Sarmatoiulus kessleri* Lohm. and its removal in the excreta. *Doklady̆ (Proceedings) of the Academy of Sciences of the U.S.S.R. (Biological Sciences),* **278,** 557–9.

Zulka, K. P. (1990). Myriapods from a central European river floodplain. *Proceedings of the 8th International Congress of Myriapodology*—Abstracts. *Veröffentlichungen der Universitat Innsbruck,* **177**.

Subject index

Page, figure and/or table numbers in italics refer to the positions in the text where terms are defined.

Systematic index

References to figures are in bold, those for tables are in italics.